The Craft of Scientific Presentations

Springer
New York
Berlin
Heidelberg
Hong Kong
London
Milan
Paris
Tokyo

The Craft of Scientific Presentations

Critical Steps to Succeed and Critical Errors to Avoid

Michael Alley

With 41 Illustrations

Springer

Michael Alley
Mechanical Engineering Department
Virginia Tech
Blacksburg, VA 24061
USA
alley@vt.edu

Cover photographs: (Top): Richard Feynman, Nobel prize winner in physics, lecturing on quantum mechanics (courtesy of the Archives, California Institute of Technology, photo 1.10-118). In this photo, Feynman demonstrates the value of communicating with gestures. Gestures and other aspects of delivery are discussed in Chapter 5. (Bottom left): Lightning demonstration at the Deutsches Museum in Munich, Germany (courtesy of the Deutsches Museum). In this demonstration, a lightning bolt strikes a church that is not well grounded. Because the church is not well grounded, a second stroke occurs between the church and a nearby house. Demonstrations and other visual aids are discussed in Chapter 4. (Bottom right): Poster presentation of capstone design projects at Pennsylvania State University (courtesy of the Learning Factory, Pennsylvania State University, 2001). The design of posters is discussed in Appendix B.

Color versions of all slides in this book can be found at the following Web site:
http://www.me.vt.edu/writing/
Ancillary information for this book can be found through the publisher's Web site:
http://www.springer-ny.com

Library of Congress Cataloging-in-Publication Data
Alley, Michael.
The craft of scientific presentations : critical steps to succeed and critical errors to avoid / Michael Alley.
p. cm.
Includes bibliographical references and indexes.
ISBN 0-387-95555-0 (pbk. : alk. paper)
1. Communications in science. 2. Communication of technical information. 3. Lectures and lecturing. I. Title.
Q223.A38 2003
808′.0665—dc21 2002030237

ISBN 0-387-95555-0 Printed on acid-free paper.

Printed in the United States of America.

9 8 7 6 5 4 3 2 1 SPIN 10887446

Typesetting: Photocomposed copy produced using PageMaker 6.5 files for the PC, prepared by the author.

www.springer-ny.com

Springer-Verlag New York Berlin Heidelberg
A member of BertelsmannSpringer Science+Business Media GmbH

◆

For two women of science—
Peggy White Alley
and
Karen Ann Thole

◆

Preface

On March 21, 1949, I attended a lecture given by Linus Pauling.... That talk was the best talk by anyone on any subject that I had ever heard.... The talk was more than a talk to me. It filled me with a desire of my own to become a speaker.[1]

—Issac Asimov

At the first stop of a tour in Japan, Albert Einstein gave a scientific presentation that, with the accompanying translation, lasted four hours. Although his audience appeared to be attentive the entire time, Einstein worried about their comfort and decided to pare back the presentation for the next stop on his tour. At the end of the second presentation, which lasted two and a half hours, the crowd did an unusual thing in Japanese culture, particularly in that era. They complained. For Einstein, though, the complaint was a compliment—this crowd had wanted him to deliver the longer version.[2]

When was the last time that you sat through two and a half hours of a scientific presentation and wished that it would go longer? Unfortunately, such responses to scientific presentations are rare. Granted, Einstein was a brilliant scientist, but just because one is a brilliant scientist or engineer does not mean that one is an engaging presenter. Consider Niels Bohr, the great physicist who won a Nobel Prize for his proposed structure of the hydrogen atom. Despite being an inspiration for many physicists,[3] Bohr had difficulty communicating to less-technical audiences. For example, his open series of lectures in the Boston area drew progressively fewer and fewer attendees because "the microphone was erratic, Bohr's aspirated and sibilant diction mostly incomprehensible, and his thoughts too intricately evolved even for those who could hear."[4]

So what is needed to become an excellent scientific presenter? This question is difficult to answer, because the presentation styles of excellent scientific presenters vary so much. For instance, Albert Einstein was humble and soft-spoken in his delivery, while Linus Pauling's delivery was dynamic and charismatic. Just because different presentation styles achieve success does not mean that any style is acceptable. For every exceptional scientific presenter such as Einstein or Pauling, ten weak presenters make their way to the podium to bore, confuse, or exasperate their audiences.

One failing that many weak presenters share is that they present their results without preparing the audience enough for those results. What occurs then is that the audience does not understand or fully appreciate what has been presented. Another common failing is that many presenters show a host of slides that follow the defaults of Microsoft's PowerPoint program, but that do not serve the audience or the situation. For instance, many slides shown at conferences contain mind-numbing lists and distracting backgrounds, but do not contain well-worded headlines or key images that would orient the audience to the work.

So how should scientists and engineers present their work? Given the diversity of audiences, occasions, and topics, establishing a set of rules for how to give a strong scientific presentation is difficult. For that reason, most rules that do exist, such as *tell them what you're going to tell them, tell them, and then tell them what you told them,* have exceptions. For instance, this often quoted strategy does not fare well with an audience that is strongly biased against the results.

Rather than present a list of simplistic rules, this book examines the styles of successful scientific presenters. Included as models are Ludwig Boltzmann, Albert Einstein, Richard Feynman, Rita Levi-Montalcini, and Linus Pauling. In addition, the book presents the experiences of other scientific presenters, such as Heinrich Hertz, J. Robert Oppenheimer, and Chien-Shiung Wu, whose initial presentations were weak, but who became strong presenters later in their careers. Moreover, the book looks at a third category of presenters, who because of obstacles never gave great presentations, but did rise above those obstacles to make successful presentations. Heading this category is Marie Curie, who overcame stage fright, hostile audiences, and her husband's tragic death, to communicate her work.

In addition to examining successes, this book considers what causes so many scientific presentations to flounder. To this end, this book considers ten critical errors that undermine scientific presentations at conferences, lectures, and business meetings. Some errors such as a speaker losing composure (Error 10) are weaknesses that everyone recognizes as errors. Other errors, such as displaying slides that no one remembers (Error 6), are such common practice that many presenters mistakenly assume that no alternatives exist.

By showing you the differences between strong and weak presentations and by identifying, for you, the errors that presenters typically make, this book places you in a position to improve your own presentations. The ultimate goal of this book is much higher than simply instructing you in how to present your work successfully. This book's goal is to give you enough insight that you can effectively critique, reflect on, and learn from your own presentations until they become outstanding.

Acknowledgments

Many scientists, engineers, and technical professionals have contributed to this book. Of particular help have been the book's reviewers: Professor Harry Robertshaw from Virginia Tech; Christene Moore from the University of Texas; Dr. Joanne Lax from Purdue University; Dr. Tom von Foerster from Springer-Verlag; and Dr. Clyde Alley from Mason-Hanger.

For their stories and insights, I must give special thanks to the following engineers, scientists, and managers: Professor Kenneth Ball from the University of Texas; Scott Dorner from OPS Systems; Bob Forrester of the United States Army; Mike Gerhard from Lawrence Livermore Lab; Professor Dan Inman from Virginia Tech; Dr. Tom McGlamery from the University of Wisconsin; Professor Patrick McMurtry from the University of Utah; and Patricia N. Smith of Sandia National Laboratories.

Finally, I must thank my students from Virginia Tech, the University of Texas, the University of Wisconsin, and the University of Barcelona. The insights, stories, and criticisms of these individuals have broadened this book's vision and deepened its advice.

Contents

Chapter 1

Introduction

It was very long ago when Richard Feynman had felt nervous at having to give a seminar.... Since then he had developed into an accomplished and inspiring teacher and lecturer, who gave virtuoso performances full of showmanship, humor, with his own inimitable brilliance, style, and manner.[1]

—Jagdish Mehra

In terms of hours spent, scientific presentations are costly. Even for informal presentations given on site, the audience members have to devote valuable time to attend, and the speakers have to give up valuable time to prepare and deliver. For presentations that require travel, the costs rise dramatically. Each year, large institutions, such as Los Alamos National Laboratory, spend millions of dollars in salary and travel expenses to have their scientists and engineers attend and make presentations.

Although expensive, scientific presentations are important. Consider that the information communicated in presentations is often only a few days old, sometimes only a few hours old. Conversely, the information in a professional journal at publication is typically a few months old, and the information in a scientific book is typically a year old at publication. For some areas of science and engineering, major advances occur so often that scientists and engineers cannot afford to wait for a publication cycle to learn the latest news. For instance, at Pratt & Whitney, the principal means of communicating new information about gas turbine engines is not documents, but presentations.[2] There, laboratory and computational results from presentations are sometimes directly incorporated into new engine designs.

Being able to make a strong presentation is not only important for communicating the work, but also important for communicating one's contribution to the work. Audiences often assign credit for the work to the person who makes the presentation, even if that person presents on behalf of a team. Moreover, the stronger the presenter is, the more the credit that the audience assigns to that presenter. This relationship of the audience assigning credit based on speaking ability was clear with the discovery of the first superconductor that had a temperature above the boiling point of liquid nitrogen. To help him in his search for this superconductor, Professor Paul Chu of the University of Houston had brought in his former student, Professor Maw-Kuen Wu of the University of Alabama-Huntsville. Chu had already identified a host of compounds that offered promise to be such a superconductor and needed help testing those compounds. When Wu and his graduate student Jim Ashburn discovered that one of the compounds was a superconductor, they contacted Chu, and the three held a press conference in Houston. Chu, being the best speaker and the leader of the team, spoke at the news conference that announced the finding. Although Chu clearly acknowledged Wu and Ashburn's contribution at the news conference, the press latched onto Chu's name. In many of the newspaper and journal articles about the discovery, Chu's name was the only one mentioned.[3]

Interestingly, a similar scenario occurred a year later in the same field when Zhengzhi Sheng, a postdoctoral researcher at the University of Arkansas, discovered another superconductor at an even higher temperature. Because Sheng was not a good speaker, the department chair, Allen Hermann, spoke at the press conference. Although Hermann repeatedly acknowledged the contribution of Sheng, Hermann was the one who received most of the accolades.[4]

Given the expense and importance of scientific presentations, scientists and engineers should strive to communicate effectively and efficiently in those presentations. Also, because scientists and engineers use both presentations and documents to communicate important work, scientists and engineers should seize upon the advantages of both media. Likewise, scientists and engineers should mitigate each medium's disadvantages.

Advantages and Disadvantages of Presentations

When contemplating whether to make a scientific presentation, perhaps a good first question to ask is, Why not just write a document or post a Web page? Given the expense of scientific presentations, writing a document or posting a Web page might be a better way to deliver the information. However, presentations offer several advantages.

Perhaps the most important advantage of a presentation is that a presentation offers someone on stage to answer questions for the audience. Answers to questions can provide the audience both with more depth about an aspect of the topic and with additional information outside the topic's original scope. In a document, the author imagines the audience and, based on that imagination, presents the topics that he or she thinks that audience needs at the levels that the audience needs. In a presentation, though, the audience can essentially revise the original presentation by requesting more depth or a broader scope.

A second advantage of making a presentation is that a presentation allows the speaker the opportunity to observe the reactions of the audience and revise the presentation on the spot for that audience. For example, during a presentation to some mathematicians, Patrick McMurtry, an engineering professor from the University of Utah, noticed from the blank looks of his listeners that they did not understand the term "laminar steady-state flow." McMurtry asked to borrow someone's lighter, clicked it on, and gave the audience an example. The smoke just above the flame rose in distinct streamlines—such a flow was laminar. However, well above the flame, these streamlines of smoke overlapped in random turns and curls—such a flow was turbulent. Because understanding the difference between laminar flow and turbulent flow was crucial to understanding the work, McMurtry salvaged the presentation with this on-the-spot revision.[5]

A third advantage of making a presentation is that a presentation offers more ways of emphasizing key points than a document does. In a document, an author can emphasize key points with repetition and placement. In a presentation, though, the presenter has all those options and one more: delivery. For instance, a speaker can pause before an important point. Also, for effect, a speaker can speak more loudly or reduce the voice to a whisper. Moreover, a speaker can provide additional emphasis by gesturing or moving closer to the audience.

So far, the advantages of a presentation have centered on the speaker's interaction with the audience. A different type of advantage of making a presentation concerns the visuals aids that one can use in a presentation. Essentially, a document is limited to an illustration that fits on a page. However, a presentation can incorporate not only the still images of a document, but also the sequential images of a film. Moreover, a presentation can incorporate color into those images more easily and less expensively than a document can. In addition, the presenter can include demonstrations. Demonstrations not only allow the audience to see the work, but also can allow the audience to hear, touch, smell, and even taste the work.

A fifth advantage of a presentation is of a legal nature. With some presentations, such as the evacuation procedures for a tall building, the presenter might want to ensure that the audience has witnessed the information. For this example presentation, the presenter can have the audience sign in when entering the room. This arrangement has advantages over a document, which might lie unopened, or a Web page, which might not be accessed.

Perhaps a better way to view the advantages of presentations is to imagine a world in which they do not exist. Such was the world of Lise Meitner when she worked at Berlin's chemistry institute in the early part of the twentieth century. Because of rules forbidding women to participate, she was not allowed to attend the chemistry seminars. Meitner, who later helped discover nuclear fission, so wanted to learn chemistry that she sometimes sneaked upstairs into the institute's amphitheater and hid among the tiers of seats to listen.[6] Almost thirty years later in the century, a similar situation existed at Oxford for Dorothy Crowfoot Hodgkin, who later won a Nobel Prize in Chemistry for discovering the structure of insulin. The chemistry club at Oxford did not permit women, even if they were on the faculty, to attend meetings. Unable to interact with others in this way, Hodgkin had difficulty attracting students until a student organization decided to invite her to speak.[7]

Although presentations have several advantages over documents, they also have several disadvantages, as shown in Table 1-1. For instance, one disadvantage of presentations in relation to documents or Web pages is that while you have the opportunity to revise a document or Web page, you have only one chance to say things correctly in a presentation. Simply forgetting a word from a sentence in a presentation can trip an audience, especially if that word is important—the word "not," for example. Likewise, in a presentation, your audience has only one chance to hear what you say. If the presentation triggers an idea for someone in the audience and that someone contemplates that idea for a moment during the presentation, then that person misses what the speaker has said. A document or Web page, on the other hand, allows readers the chance to review a passage as many times as they desire.

A second disadvantage of a presentation is that the audience has no chance to look up background information. If in a presentation the speaker uses an unfamiliar word, such as "remanence," and does not define the word, then the audience is stuck. If the presentation's format does not allow for questions until the presentation's end, then members of the audience sit frustrated wondering what "remanence" means. With a document or Web page, though, the reader has the chance to look up "remanence"

Table 1-1. Advantages and disadvantages of making a presentation.

Advantages	Disadvantages
Opportunity to receive and answer questions	One chance for speaker to talk; one chance for audience to hear
Opportunity to revise on the spot	No chance for audience to look up background information
Opportunity to use delivery for emphasis	Audience restricted to pace of speaker
Ability to incorporate many types of visual aids	Success dependent on speaker's ability to deliver
Assurance that audience has witnessed the information	Difficulty in assembling speaker and entire audience at one time

(which is the residual magnetic flux density in a substance when the magnetic field strength returns to zero).

Yet a third disadvantage of a presentation is that the audience is captive to the pace of the speaker. Unlike the pace of a document, in which an audience can read as slowly as is needed for understanding, the pace of a presentation is determined by the speaker.

The fourth disadvantage, which can be an advantage depending upon the speaker, is that the success of the presentation depends upon the delivery of the speaker. If the speaker is so nervous or befuddled that he or she cannot communicate the ideas to the audience, the presentation will not succeed. Delivery can cause a big swing in the perception of a presentation. Some speakers, such as Linus Pauling, had the charisma to make a presentation stronger than perhaps it actually was. Other speakers, such as Niels Bohr, undermined their content with a delivery that distracted or prevented the audience from understanding the message.

A final disadvantage of presentations is one of timing: how to gather everyone at a particular time to attend the presentation. Granted, teleconferencing can overcome this problem, but not everyone can afford this solution. Videotaping is a less expensive alternative, but videotaping loses one of the main advantages of presentations, namely, the interaction with the audience. Another issue with timing is the attention span of the audience. Although some people can listen attentively for more than one hour, many people become tired and restless after only twenty minutes. When the technical subject is deep and complex, the task of communicating that subject solely with a presentation becomes difficult.

So far, this discussion has centered on the effects of presentations upon the audience? What about the effects of presentations on the speaker? As with writing scientific documents, making scientific presentations can help solidify one's ideas. On a number of occasions, when I have presented an idea that I had spoken about scores of times before, I suddenly found a new and interesting perspective on those ideas. Perhaps the discovery arose from a question posed by the participants or from a different order of presenting details. Whatever the source was, I now saw the subject at a deeper level and could present the idea more clearly than I had ever presented it.

Richard P. Feynman claimed to have experienced such moments of discovery.[9] So has my wife. While she was preparing for one review meeting, a question had been dogging her: How can I sell this company on the measurement technique that I want to research? Her technique was a new, but time-consuming, way to measure the heat transfer on some leading edges in a flow, a technique that she felt would be a difficult sell, given the project's time constraints. Then, as she was walking up to speak, the idea struck her that she would simply ask the company sponsors for some of their new heat exchangers on which to try the measurement technique. The sponsors would receive valuable information on their own product, and she would receive funding for her research. On the fly, she proposed these measurements, and the company sponsors became quite excited about the prospect. Although she was nervous about proposing something that she had not spent time thinking through, the idea had felt so right that she had gone with it.

Four Perspectives on Presentations

Given the advantages and disadvantages of presentations, this book attempts to offer advice that emphasizes the advantages, while mitigating the disadvantages. In doing so, this book analyzes presentations from four perspectives. The first perspective is speech, which encompasses the words that you say. The second perspective is structure, which is the organization, depth, emphasis, and transition between major points. Third is the perspective of visual aids. In this book, visual aids include not only projected slides, posters, models, and writing boards, but also films and demonstrations. The final perspective is delivery, which is one's interaction with the audience and the room. For a summary of the advice from these four perspectives, see Appendix A.

In presenting these four perspectives, this book anchors its advice with scores of examples gathered from conferences, symposiums, and business meetings. In essence, this book pursues a similar study to the one that Michael Faraday undertook as a young scientist when he examined the different styles of presenters.[9] As with Faraday's study, this book's study seeks to determine what makes one scientific presentation strong and what makes another weak.

Many of the examples chosen are from famous scientists and engineers. Some of these scientists and engineers are considered excellent presenters, while others are not. Certainly, such characterizations are inherently imprecise. For one thing, not everyone is an excellent presenter every single day; in a career, everyone is likely to have a few weak presentations or at least a few presentations that are not well received. Also, some individuals, such as Maria Goeppert Mayer, were excellent presenters in front of colleagues and friends, but shy and stiff in front of strangers.[10] Moreover, not everyone is in agreement about who was an excellent presenter and who was not. For example, the opinions about the presentation skills of the engineer Willard Gibbs varied widely.[11] That variety of opinions about the effectiveness of a presenter is not surprising; to see this spread, one simply has to read a set of teaching evaluations of a university professor.

Implicit in the opinions held by an audience of a presentation are the biases of the audience toward the subject and speaker. An example of this point is James Watson's criticism of the presentation given by x-ray crystallographer Rosalind Franklin in November 1951. Watson, who with Francis Crick is credited with the discovery of the structure of DNA, criticized Franklin for her "quick, nervous style" and her lack of "warmth."[12] Although Watson restricted his comments to the delivery of her presentation, what became clear later on was the influence that her x-ray diffraction work exerted on his own thinking about the structure of DNA. Moreover, any assessment of Franklin's delivery should have accounted for the stress that she was under during this presentation. Her supposed collaborator at King's College, Maurice Wilkins, was one of the fifteen people in attendance at the colloquium. According to Watson, Maurice Wilkins wanted Franklin to work as his assistant rather than to do independent research.[13] In such a situation in which the speaker senses such tension from someone in the audience, delivering the warm and relaxed presentation that Watson apparently desired was out of the question.

Although the circumstances and variety of opinions by the audience make it difficult to draw conclusions about the effectiveness of many historical presentations such as Franklin's, the effectiveness of other past presentations is clear. For instance, Richard Feynman's lecture series on freshman physics at Caltech received so many glowing reviews and had such a profound effect on so many people that this series was undoubtedly a brilliant success.

While analyzing presentations from these four perspectives offers advantages, such discussions can skew the overall effect of a presentation. After all, a presentation that has weak slides might be strong enough in the delivery that the overall effect is positive. Still, if any of these areas is so weak that it distracts the audience from the content of the presentation, then the presentation has not reached its potential.

One perspective of presentations not considered in this book is content. An assumption for all the advice in this book is that the technical content of the presentation is worthwhile. Otherwise, it does not matter how well designed the projected slides are or how smooth the delivery is: The presentation is doomed.

Interestingly, in science and engineering there exists a deep-seated distrust of a noticeable style, what many refer to as "glitz." Certainly, style without content reduces to entertainment. If you are going to dazzle the audience in a scientific presentation, you should do so with your content (your ideas, findings, and conclusions) rather than with your style (the way that you present that content). However, that is not to say that style is unimportant; quite the contrary. Style is the vehicle for communicating the content. Presentations without attention to style often leave little of value in their wake. Granted, the content has been presented, but not in such a way that the audience understands it or realizes its importance. Strong presentations require both content and style. Content without style goes unnoticed, and style without content has no meaning.

Chapter 2

Speech: The Words You Say

Desperately eager to reach his students, his sensitivities sharpened by his own past difficulties, Oppenheimer made it a point to pay as much attention to the troubles of his charges as to the intricacies of his subject. His language evolved into an oddly eloquent mixture of erudite phrases and pithy slang, and he learned to exploit the extraordinary talent for elucidating complex technical matters. [1]

—Daniel J. Kelves

Simply put, speech is what you say in a presentation. A speech targeted to the audience is essential for a presentation's success. Consider J. Robert Oppenheimer's early lectures given at California-Berkeley in 1929. Only twenty-five years old, but already well known for his work on the quantum theory, Oppenheimer began his teaching that first semester with a class full of eager graduate students. Halfway through the semester, though, the number of students registered for his course had dropped to one.[2]

The principal reason that students dropped the course was that Oppenheimer did not target his speech to them. For one thing, Oppenheimer's pace was much too fast for the students. Interestingly, although the students considered the pace to be much too fast, Oppenheimer felt that it was too slow.[3] Another problem with Oppenheimer's speech was that he made "obscure references to the classics of literature and philosophy."[4] The combination of these two problems caused many of the students to complain to the head of the department. However, Oppenheimer was already aware of the problems and worked hard to slow his pace, to clarify his ideas, and to make connections between his points. The result was that Oppenheimer's later students found him to be "the most stimulating lecturer they had experienced."[5]

One important element of speech that Oppenheimer failed to achieve in his early lectures was the matching of what was said to the audience, purpose, and occasion. When this match does not occur, one essentially gives the wrong speech. Another important aspect of speech with which many young scientists and engineers struggle involves the source of words for the speech. Do the words arise extemporaneously, from memory, from reading, or from points (which may be memorized or written on note cards or presentation slides)? The occasion of the presentation dictates which of these sources should be used, and many times when the wrong source is chosen, the presentation fails. Before examining these two critical errors of speech, this chapter discusses different ways for making one's speech distinct and different ways for supporting arguments within speech.

Adding Flavors to Your Speech

Rather than simply presenting the work in a dry manner, the best speakers flavor their speeches. One such flavor is the incorporation of analogies, examples, and stories. Another flavor is achieving a personal connection with the audience. Still another flavor is to bring in humor. Not only do these flavors give individuality to one's presentation, but they also serve the audience. For instance, analogies, examples, and stories serve as mnemonics when the audience tries to recount the presentation. In addition, personal touches engage the audience, and humor allows the audience to relax and participate.

Incorporating Analogies, Examples, and Stories. When you want to make a segment of your presentation memorable, then consider using analogies, examples, or stories.

For instance, when the purpose of a portion of a presentation is simply to convey the size of something or the likelihood of an event, analogies are powerful. For instance, Otto Frisch liked to use the following example to describe the size of a nucleus: "If an atom were enlarged to the size of a bus, the nucleus would be like the dot on this *i*."[6] Einstein used the analogy of "shooting sparrows in the dark"[7] to describe the likelihood of producing nuclear energy with alpha particles striking nitrogen nuclei. When describing his work with turbine blades in gas turbine engines, the engineer Fred Soechting uses the following analogy: "The amount of power produced by a single gas turbine blade equals that of a Masarati sports car."[8] Such descriptions, when they support the presentation's content, are *keepers*: things that audiences hold onto when they leave the room. Too often, I attend a presentation and a couple of days later remember nothing about that presentation: not a result, not an image, not an observation, not even a striking detail. One test for the success of a presentation is what the audience remembers two days later.

Examples are important in a different way for audiences. Often, presentations fail because the speaker restricts the speech to an abstract or mathematical perspective. While some people can learn from this purely mathematical perspective, most cannot. Most people require some image or physical process to follow. Consider the difference between listening to the solution of a second-order differential equation and listening to the solution of a second-order differential equation that represents the flight of a paratrooper dropped from a plane. In the second presentation, you have something physical to which you can anchor the mathematics. When listening to presentations of mathematical derivations, Richard Feynman would request physical examples for the equations shown. To the surprise of the presenter and everyone else in the room, Feynman would sometimes catch errors in the middle of detailed derivations because while everyone was desperately trying to follow the mathematics, Feynman was working through the physics of the example.[9]

When the speaker desires the audience to experience a project in a more personal way, stories can serve presentations. The astronaut and physicist Ellen Ochoa effectively uses stories to show audiences what it is like working on the space shuttle.[10] As president of Sandia National Laboratories, C. Paul Robinson often finds occasions to interweave stories into his presentations. For instance, in one presentation, he had an audience on the edges of their seats by recounting Sandia's efforts to verify a missile treaty.[11] Also noted for incorporating stories into their presentations were Feynman, Linus Pauling, and Albert Einstein.

In addition to allowing the audience to experience a project, stories can serve long presentations by giving the audience a needed rest break. An advantage of incorporating stories is that they are relatively easy to recall. If you live through an experience (or even hear of an experience told to you), you can usually recall the sequential points of that experience days, weeks, even years later. The powerful effect of stories is that audiences can do the same. For that reason, stories can serve as mnemonics for the audience when they try to remember points of the presentation.

Making a Personal Connection. Another flavor that many people successfully incorporate into speech is a personal connection. Michael Faraday and Ludwig Boltzmann were noted for giving presentations that had a warm and personal atmosphere. At a time when so many others spoke for the sole purpose of impressing audiences with their knowledge, Faraday worked hard to make sure that everyone in the audience understood what he had to say. His eye contact, his humbleness, his passion for having the audience understand him — these served to make connections with his audience.[12]

Ludwig Boltzmann, the developer of the statistical treatment of atoms, made his presentations personal by stating things about himself. Teaching at a time when most professors adopted a formal distance from the students, Boltzmann broke tradition and made personal connections with his audience. According to Lise Meitner,

> Boltzmann had no inhibitions whatsoever about showing his enthusiasm when he spoke, and this naturally carried his listeners along. He was fond of introducing remarks of an entirely personal character into his lectures. I particularly remember how, in describing the kinetic theory of gases, he told us how much difficulty and opposition he had encountered because he had been convinced of the real existence of atoms and how he had been attacked from the philosophical side without always understanding what the philosophers had against him.[13]

Boltzmann's personal style seemed to suit his purpose and contributed to his ability to inspire. Confirming his abilities was his legacy of pupils: Svante August Arrhenius, Paul Ehrenfest, Fritz Hasenöhrl, Stefan Mayer, Lise Meitner, and Walter Nernst.[14]

Other speakers make the speech of their presentations personal by showing connections between their own work and the work done by members of the audience. Such speakers often refer to those audience members by name during the presentation. This style can be particularly effective if you find yourself having to explain something to an audience that includes an expert who knows much more than you do about a topic in your talk. For instance, my wife is primarily an experimentalist, but uses commercial computational codes such as Fluent in her work. When she gives a conference presentation, she usually has prominent computationalists in her audience. Given that, in explaining the principles of her commercial code, she respectfully acknowledges the computationalists who could explain the code better than she can, and then she explains the code as well as she can. Naming those computationalists during the presentation not only serves as a sign of respect, but also recruits them to her side.

In teaching large classes, one of the best ways to make a personal connection is to do the unexpected and to learn who the audience is. At Virginia Tech, Professor Harry Robertshaw and I teach a two-semester measurements course to more than two hundred mechanical engineering students. Using a technique of Professor Wallace Fowler from the University of Texas, we photograph the students at the beginning of the first semester so that we can learn their names. In addition, we survey the students to learn what measurements they have done in their co-ops and summer jobs. Whenever possible in our lectures, we then mention the experiences of individual students. Because of these efforts, presentations that the students assumed were going to be anonymous experiences have become personal experiences.

One effect of our efforts has been that the students concentrate more during the lectures. That effect we expected, because the students now have to be prepared for us calling upon them by name at any moment. An unexpected result, though, has been that the students have put much more stock into what we have to say. In other words, our making a personal connection to the audience has increased our credibility with that audience.

Incorporating Humor. For his series of Messenger Lectures at Cornell, Richard Feynman was introduced as someone who had won the Albert Einstein Award in 1954, who had served on the Manhattan Project during the Second World War, and who played the bongo drums. Feynman began his lecture with the following statement: "It is odd, but on those infrequent occasions when I have been called upon in a formal place to play the bongo drums, the introducer never seems to find it necessary that I also do theoretical physics."[15]

In one of his presentations as president of Sandia National Laboratories, C. Paul Robinson began in the following way: "As a small boy I had two dreams, and I was torn between them. At times I wanted to become a scientist, and at other times I just wanted to run away and join the circus. But thanks to the grace of God and a career in the Department of Energy's laboratories, I've been able to fulfill both dreams."[16]

Humor can relax an audience. Humor can also allow an audience to respond to a presentation. Moreover, humor can engage an audience and can give an audience a needed rest. However, because attempts at humor are risky, several points about humor are worth noting. First, not everyone is suited to make a crowd laugh. Granted, humor comes in various forms: Some people's humor is dry, and other people's humor is dramatic. Although these different ways exist to make people laugh, not everyone's attempt will work. In fact, my experience is that most people who try to be funny in a professional situation, especially before an audience whom they do not know, draw more groans than genuine laughter.

A second point is that although some books on presentations suggest that the speaker should open each presentation with a humorous remark, the beginning of a talk is probably the most difficult time to make people laugh, especially if those people do not know the speaker. One reason is that humor usually arises from saying something that no one expects, but that contains some truth. The unexpected realization of truth then makes people laugh. At the beginning of a presentation in which people do not know the speaker, the audience does not know what to expect of the speaker. Another reason that opening the presentation with humor is difficult is that the speaker is usually the most nervous then. Moreover, if the remark fails to draw warm laughter, the speaker could easily become more nervous. Worse yet, a failed attempt at humor at the beginning could cause the audience to feel ill at ease with the speaker, and the beginning of a presentation is when the audience makes perhaps its most important assessment of the speaker. For these reasons, I believe that it is far more effective to wait until the middle of the presentation, when the speaker has developed credibility with the audience and when the remark will truly be unexpected.

A third point is that humor is risky in a professional situation. What might strike people as funny in a restaurant during an informal lunch can come across as crass in a formal meeting where the audience members are seated with their managers and colleagues. Moreover, if the speaker touches on a controversial subject, humor can irritate an audience. What subjects risk controversy in a professional setting? Certainly, comments about sex are taboo, because some managers and colleagues impose an atmosphere of sexual tension in the workplace. Some people claim that such comments would have been acceptable thirty years ago (the supposed "good old days"), but the truth is that they were not. The same uncomfortable situations existed then; it was just that the discomfort of those situations had not been exposed.

Defining the line of what will make everyone laugh and what will make some people feel uncomfortable is impossible. People react differently to different subjects at different times in their lives. Just remember that in any large professional crowd, someone is probably sensitive to race, gender, religion, or death. So what topics are appropriate? Typically, stories about your own failings are the safest. Professor Dan Inman, a vibrations engineer, is well known for the humor that he works into his talks. For that reason, he is often asked to give after-dinner talks at conferences. Inman believes that self-effacing humor is best. "I'm considered funny because I'm such an easy target," he says.[17] In addition, Inman believes that humor should be natural, not planned. Moreover, he feels that humor is not appropriate for every situation. If his first attempt at humor does not elicit laughter, then he backs off and plays the situation straight. Finally, Inman notes that a problem with continually using humor is that people continually try to read funny things into what he says, even when he is serious.

Supporting Arguments in Your Speech

In addition to the different styles that presenters have in their speech, presenters incorporate different types of evidence to support the assertions of the speech. According to Aristotle, this evidence falls into three categories: appeals to logic, appeals to the emotion of the audience, and appeals to your own character. If asked which of these categories exerts the greatest influence on them, most engineers and scientists would name appeals to logic. While most scientists and engineers would say that appeals to logic influence their decisions the most, the appeals to character and emotion play more important roles than most scientists and engineers realize. Moreover, many political decisions about science and engineering are not made by engineers and scientists. Rather, politicians make these decisions, and these individuals often are swayed by appeals to character and emotions. For that reason, understanding the influence of these different appeals is important.

Appeals to Logic. Logical evidence varies from deductive and inductive reasoning to statistics, referenced findings, examples, and analogies. Not all of these have the same level of strength, as suggested by the ranking in Table 2-1. For instance, deductive reasoning and inductive reasoning are the most powerful, while analogies when used to propel arguments usually follow the axiom as being the "weakest form of argument."

Of the different types of logical evidence, deductive reasoning is considered the strongest. Deductive rea-

Table 2-1. Different types of logical evidence in descending order of strength.

Type of Evidence	Example
Deductive Reasoning	Mammals are all creatures that are warm-blooded and breathe oxygen; whales are warm-blooded and breathe oxygen; therefore, whales are mammals.
Inductive Reasoning	The gravitational force
Referenced Facts	The combustion gases in a gas turbine engine reach temperatures more than 500°C hotter than the melting temperature of the steel.[18]
Statistics	Reducing the temperature on a gas turbine blade from 1140K to 1090K increases the blade's life from 560 hours to 3900 hours.[19]
Examples	Earthquakes can cause many deaths. For example, the 1976 earthquake in Tianjin, China, killed more than 242,000.[20]
Analogies	Just as the designs for atomic bombs were reduced from the bulky size in Fat Man to the size of a soccer ball within a decade, so too could designs of neutron bombs, making them extremely dangerous as tools for terrorists.[21]

soning usually takes the form of a syllogism: Given A and given B, then C follows. A good example of how deductive reasoning influenced a persuasive presentation occurred in the decision by the United States Congress on where to place the superconducting supercollider, which was to be a huge particle accelerator. Because this experiment was to create hundreds of jobs and bring millions of dollars into the local area, more than forty-three proposals were submitted for the site. Ellis County, Texas, which won the contract, used deductive reasoning in its arguments.[22] This reasoning consisted in the premise that the collider site had to meet several criteria, including relatively flat terrain, few freezing days, little seismic activity, and low rainfall. For each of these criteria, some of which are shown in Figure 2-1a, the presenters of the proposal used referenced facts and the opinions of experts to assign a cut-off value. The establishment of these criteria formed the A-portion of the syllogism. Then with a map of the United States, the presenters used overlays as shown in Figure 2-1b to shade those parts of the country that did not meet the stated criteria. This application of the criteria to the map constituted the B-portion of the syllogism. When all the overlays had been placed upon the map, only one small circle in Ellis County, Texas, remained without shading, as shown in Figure 2-1c. That statement became the C-portion of the syllogism and the main evidence that contributed to the awarding of the contract.

Statistics are another form of logical evidence, and their power varies widely. At the more persuasive end are experimental data that show definite trends. At the weaker end is the comparison of data that are not comparable. An often quoted statistic concerns the amount of research funding from the National Institutes of Health (NIH) that has gone to fight the AIDS epidemic. In 1998, for example, NIH distributed $2400 per patient in research

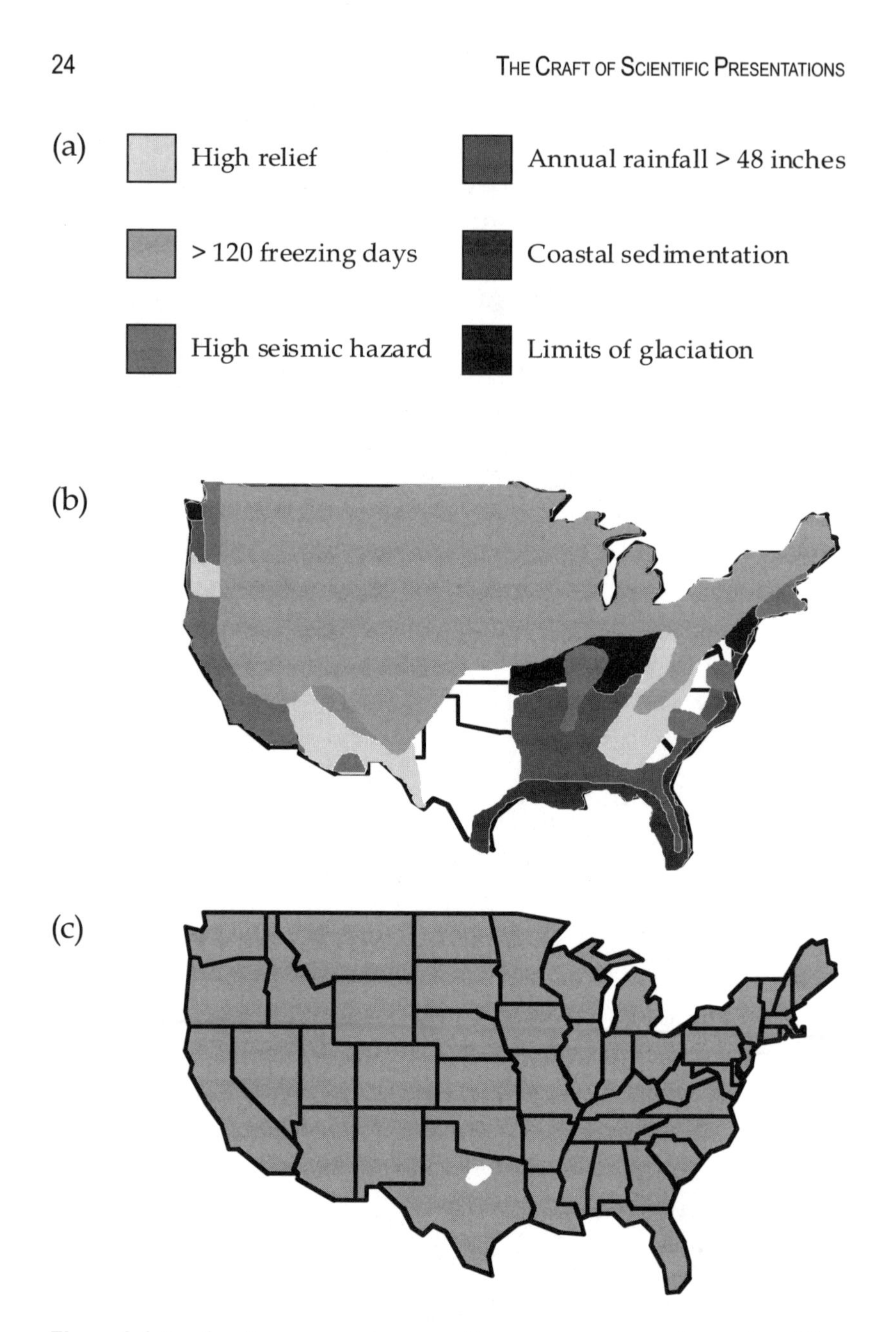

Figure 2-1. Deductive reasoning used by presenters to show that Ellis County, Texas, was the best site for the superconducting supercollider.[22] The reasoning involved first establishing the site criteria, some of which are listed in (a). Then, as shown in (b), those criteria were applied in overlays to a map of the continental United States. As shown in (c), only one area of the country, Ellis County, satisfied all the criteria.

funds to fight AIDS, which was the number-17 killer in the country that year, but spent only $108 per patient to fight heart disease, which was by far the number-one killer in the country that year.[23] The statistic suggests that too much money is being spent on fighting AIDS. That assertion might very well be valid, but the statistic does not account for all variables: how recently AIDS was discovered, how quickly the number of deaths from AIDS has risen, the severity of prognosis for AIDS in terms of life expectancy, or how much progress in fighting AIDS those research dollars have produced.

As with the power of statistics, the power of examples varies dramatically. The power of an example depends upon the assertion that it is to support. For instance, to support the argument that a drug is dangerous, a single example of someone who was harmed by the drug can be powerful. However, to support the argument that a drug is safe, a single example of someone who used the drug with no side effects does not carry nearly as much weight.

Although useful for explaining how things work or how large things are, analogies are generally not effective for supporting assertions in an argument. The reason is that because analogies compare two dissimilar things from one perspective, a skeptical audience can easily point out differences between those two things that would affect that comparison.

Appeals to Emotion. While all scientists and engineers agree that appeals to logic are important in an argument, many scientists and engineers underestimate the importance of appeals to emotion, especially when the audience making the decision is nontechnical. For instance, greatly influencing the political decision to stop building nuclear power plants in the 1980s was the appeal to the emotion of fear made by antinuclear groups. Although the nuclear power industry countered with logical evidence such as the statistic that coal plants emit far more radiation than the typical nuclear power plant, the appeal to fear by the antinuclear groups had the larger influence.

Numerous examples exist in which appeals to emotion significantly influenced decisions: protecting endangered wildlife, protecting forests and rivers, and increasing the research funds to fight a disease. As mentioned, an interesting case has been the amount of research funding from the National Institutes of Health (NIH) that has gone to fight the AIDS epidemic. Certainly, the relatively recent discovery of AIDS and its rapid increase in cases account for much of this funding, but also contributing have been the emotional and widely publicized appeals for research funding from AIDS activists.

Appeals to Character. An appeal to the character of the speaker can have a deep influence in a persuasive presentation. If a relatively unknown scientist suggests that Vitamin C is the most important vitamin for a long and healthy life, that suggestion does not receive nearly as much attention as when Linus Pauling, a Nobel Prize winner, makes the same suggestion. Likewise, when some relatively unknown engineer expresses concern that the size of a neutron bomb could be reduced to the dimensions of a soccer ball (thus making it an extremely dangerous weapon for terrorists), that statement does not carry nearly as much weight as when it is made by Sam Cohen, the inventor of the neutron bomb.[24]

Even those of us without such credentials can use this type of appeal effectively, such as when we adopt a position that is counter to our background or history. For instance, Walter Mossberg's argument a few years ago against a proposed operating system for Apple's Macintosh computer began with the point that he did not relish taking that position. After all, over the years, he had been a staunch Macintosh supporter and was quoted widely in many of Apple's advertisements. However, he felt that the released operating system demanded too much faith on the part of Macintosh's users.[25]

Character includes your reputation with audiences. Chien-Shiung Wu, the physicist who performed the first experiment showing that nuclear particles violate the law of parity, earned a reputation as a physicist whose work was to be trusted.[26] Such reputations come only after hard work and many tests. When Wu found a result that did not agree with the results of someone else, she did not end her argument by simply showing that her results were correct. She also worked to show why the other results were incorrect.

Character also includes your connection to the audience. As mentioned, in their speech, Michael Faraday and Ludwig Boltzmann made personal connections to their audiences. These personal connections were appeals to character that were designed to earn respect. Faraday believed that for a speaker to be effective, the audience must like and trust the speaker.[27] To achieve that respect, Faraday believed that the speaker should first respect the audience. Boltzmann held that same respect for his audience. According to Fritz Hasenöhrl, who was a student of Boltzmann, "[Boltzmann] never exhibited his superiority. Anybody was free to put questions to him and even to criticize him. The conversation took place quietly and the student was treated as a peer. Only later one realized how much he had learned from him."[28]

Critical Error 1
Giving the Wrong Speech

> *Rutherford, though always inspiring, was not a great lecturer – "To 'Er' was Rutherford!" Bohr was much worse. His failing was that he used too many words to express any idea, wandering about as he spoke, often inaudibly.*[1]
>
> – Sir Mark Oliphant

On January 27, 1986, because of the low temperatures expected for the next morning's launch of the space shuttle *Challenger*, engineers at Morton Thiokol requested a delay in the launch. From their examinations of previous launches, the engineers knew that the lower the launch temperature, the more likely that explosive gases from the solid booster rockets would escape. In an afternoon meeting, these engineers succeeded in persuading management at Morton Thiokol to request a delay. However, when Morton Thiokol's engineers and management discussed the delay with NASA during a teleconference that evening, they met strong resistance.[2] After spending almost two hours in a conference call and reviewing thirteen presentation slides faxed from Morton Thiokol, NASA remained unconvinced of the danger. Moreover, NASA's opposition to the delay was so adamant that Morton Thiokol's management rescinded the request.

The next day, sixty-three seconds into the launch, a solid rocket booster of the space shuttle *Challenger* exploded, killing all seven crew members on board.

One reason that Morton Thiokol's presentation failed to persuade NASA was that the presentation did not target the audience. For instance, in their presentation, the engineers and management at Morton Thiokol did not anticipate the strong bias that NASA had against delaying the launch. NASA had already delayed the launch more than once and felt pressure to place *Challenger* into orbit.[3]

Not targeting the audience is one common reason for the failure of many scientific presentations. Another common reason is a failure to understand the purpose of the presentation. Few presentations have the sole purpose of informing. Most scientific presentations, such as the Morton Thiokol presentation, must persuade audiences. Other presentations, such as a lecture in a university class, call for inspiring the audience.

Yet a third reason that many scientific presentations fail is that the speaker has not carefully considered the occasion of the presentation. Occasions vary greatly, from informal meetings to formal symposiums. The occasion affects the expectations that the audience has for the presentation. For instance, an audience for a morning plenary session at a formal conference has much different expectations from what an audience for an after-dinner talk has.

Targeting the Audience

Morton Thiokol's presentation to NASA provides a clear example of targeting the wrong audience. Given in Figure 2-2 are the first two presentation slides that Morton Thiokol faxed to NASA. The second slide contains data that supposedly state and support the main assertion of the presentation, namely, that the lower the launch temperature, the more erosion that the O-rings of the solid rocket boosters would likely incur and thus the more likely that explosive gases from the rocket would escape.

The information on the second slide is inappropriate for someone with a general scientific background. As Edward Tufte points out in his book *Visual Explanations,*[4]

Weak Slide

Temperature Concern on SRM Joints

27 Jan 1986

Weak Slide

HISTORY OF O-RING DAMAGE ON SRM FIELD JOINTS

		Cross Sectional View			Top View		
	SRM No.	Erosion Depth (in.)	Perimeter Affected (deg)	Nominal Dia. (in.)	Length Of Max Erosion (in.)	Total Heat Affected Length (in.)	Clocking Location (deg)
61A LH Center Field**	22A	None	None	0.280	None	None	36° - 66°
61A LH CENTER FIELD**	22A	NONE	NONE	0.280	NONE	NONE	338° - 18°
51C LH Forward Field**	15A	0.010	154.0	0.280	4.25	5.25	163
51C RH Center Field (prim)***	15B	0.038	130.0	0.280	12.50	58.75	354
51C RH Center Field (sec)***	15B	None	45.0	0.280	None	29.50	354
410 RH Forward Field	13B	0.028	110.0	0.280	3.00	None	275
41C LH Aft Field*	11A	None	None	0.280	None	None	--
410 LH Forward Field	10A	0.040	217.0	0.280	3.00	14.50	351
STS-2 RH Aft Field	28	0.053	116.0	0.280	--	--	50

*Hot gas path detected in putty. Indication of heat on O-ring, but no damage.
**Soot behind primary O-ring.
***Soot behind primary O-ring, heat affected secondary O-ring.

Clocking rotation of leak check port - 0 deg.

OTHER SRM-15 FIELD JOINTS HAD NO BLOWHOLES IN PUTTY AND NO SOOT HEAR OR BEYOND THE PRIMARY O-RING

SRM-22 FORWARD FIELD JOINT HAD PUTTY PATH TO PRIMARY 0-RING, BUT NO O-RING EROSION AND NO SOOT BLOWBY. OTHER SRM-22 FIELD JOINTS HAD NO BLOWHOLES IN PUTTY.

Figure 2-2. Reproduction of first two presentation slides from a set of thirteen that were faxed by Morton Thiokol to NASA to request launch delay of the space shuttle *Challenger* (January 27, 1986).[5] The first slide is weak because it does not establish authority: No name or company logo is given. The second slide is weak because its assertion is not explicitly stated, and the evidence for its underlying assertion is confusing.

absent from this second slide is the assertion that lower temperatures produce more damage. Also, the statistical evidence to support that assertion is buried in too much detail: the confusing names for the previous launches, the unnecessary cataloguing of the types of erosion, and the unnecessary details about the locations of the damage. Moreover, missing from this slide is key information to support the assertion, specifically, the temperatures of the different launches.

Targeting a Specific Audience. Targeting a specific audience is critical for communicating one's work. In general, the less technical the audience, the more difficult that targeting is, because the speaker has to anticipate more terms and background information that the audience needs for understanding the work. Richard Feynman understood this point. After receiving his Nobel Prize, he had been invited to Berkeley to give a lecture to an audience that he assumed would be physicists in his specific field. Upon entering the lecture room, though, he was upset to find a huge crowd of people, who were not nearly as technical as the one for which he had prepared.[6]

My father, who served for several years as plant manager of the Pantex Nuclear Weapons Facility, claims that a common way many speakers fail to target the audience is that they neglect to define their jargon. For instance, speakers will toss out abbreviations such as *HE* (which means *high explosives*) or *NC* and *NG* (the explosives *nitrocelluose* and *nitroglycerin*) without considering whether the audience knows those terms. Note that not all abbreviations are necessarily less familiar to the audience than the terms for which they stand. For instance, *TNT* is more widely known than the term *trinitrotoluene.*

Although an inspiration to many physicists including Lise Meitner and Otto Frisch, Niels Bohr was not adept at communicating to audiences who were not knowledgeable about physics. Why did Bohr struggle to communicate? Part of the problem was language; he often intermixed German and English, neither of which was his native tongue, Danish.[7] Another part was Bohr's passion for being precise. Bohr often focused on the edges of what he knew. According to Einstein, Bohr stated "his opinions like one perpetually groping and never like one who believes himself to be in possession of definite truth."[8]

Bohr's attention to precision in his speech, unfortunately, occurred at the expense of clarity. Notice in the beginning of his Nobel Prize address how his striving for accuracy causes his first sentence to lengthen to the point of being difficult to follow:

> Today, as a consequence of the great honor the Swedish Academy of Sciences has done me in awarding me this year's Nobel Prize for Physics for my work on the structure of the atom, it is my duty to give an account of the results of this work, and I think that I shall be acting in accordance with the traditions of the Nobel Foundation if I give this report in the form of a survey of the development which has taken place in the last few years within the field of physics to which this work belongs.[9]

A single sentence of this length an audience can handle, but when most of the sentences have this kind of wandering, the audience is pressed to stay with the speaker.

The above example shows what not to do in targeting an audience. Now the questions arises, How do you target a specific audience? When your audience consists of people whom you know well, targeting the audience is straightforward. As you prepare the presentation point by point, you continually ask yourself two questions: (1) Will the audience understand these points? and (2) Will the audience be interested in these points?

A more difficult situation arises when you do not know the audience well. Before such a presentation, many good speakers move out into the audience before the presentation and ask questions: What kind of work do you do? Why did you come today? What do you know about the presentation's topic? This tactic is not only important for targeting the audience, but also effective at alleviating nervousness (see Critical Error 10). In the situation of the audience not being available beforehand, many good speakers try out their presentation on someone who knows or has the same background as the intended audience.

Dan Hartley, a former vice president at Sandia National Laboratories, was one of the most adept individuals I have known at targeting an audience. While managing the Combustion Research Facility in Livermore, California, he met with many visitors, including politicians, managers from industry, Department of Energy officials, and visiting scientists from abroad. I saw him give the same tour three times in a single day, but to three different audiences. On these occasions, Hartley tailored the examples, the depth, and the background information for each group of visitors. As he spoke, Hartley constantly watched the expressions of the audiences to gather whether what he was saying registered with them.

In my own presentations, I find that thinking beforehand about the audience is helpful. In the shower or on a noontime run, I mentally work through the presentation that I am to give. In addition, during the presentation, like Hartley, I find that much can be gathered by the response of the audience. If they appear puzzled or are not making eye contact with me, then I work harder to engage them. In such instances, I often step closer to them and try to rephrase what I have just said.

After the presentation is also a fruitful time to think about the presentation. Generally, if you present a subject once, you will have to present it a second time. When reflecting on a presentation, I scroll through my slides and think about the questions raised by the audience. Perhaps those questions arose because I needed to explain certain points better in the body of the presentation. I also consider the comments that the audience made: not only what they responded to, but also what they did not respond to. For me, a presentation lasts much longer than the time I am on the stage: The planning, delivering, and reflecting usually last for days, sometimes weeks. Sometimes, long after I have given a presentation, an idea will come to me about how I could have reached the audience more effectively. These ideas I jot down in the computer file that contains my presentation slides.

Targeting Multiple Audiences. More difficult than targeting a single audience is the task of reaching a multiple audience. Barbara McClintock, who won a Nobel Prize for her work in genetics, had difficulty with this situation. McClintock communicated her work to other geneticists, but struggled to reach people outside her discipline. In fact, no one at Cornell paid much attention to McClintock's thesis work until a postdoctoral student arrived who had worked for the geneticist Thomas Hunt Morgan. This postdoctoral student not only took notice of McClintock's work, but also explained its importance to others at Cornell.[10] The result was that McClintock, still a graduate student, became the leader of a research group of postdocs.

Much later in her career, McClintock still struggled to communicate to a wider audience. In an hour-long presentation at Cold Spring Harbor in 1951, McClintock failed to communicate her work on transposons ("jumping genes") to those outside of genetics. This work, for which she eventually won a Nobel Prize more than a quarter of a century later, was dismissed by the molecular biologists at that presentation. According to Sharon Bertsch McGrayne,[11] McClintock presented the intricacies of her work in a fashion that was just too dense for this audience to digest. Disheartened by the rejection of her work, on which she had spent years, McClintock pared back efforts to communicate her work to the outside world. That she received the recognition of the Nobel Prize is a testament to how important the work was.

A multiple audience usually includes specialists in the field who understand the problems that you have faced and are interested in your designs and results. The audience also usually includes engineers and scientists from other fields. Although these engineers and scientists might understand the general theories upon which you have based your work, they may not appreciate the importance of your work. They also may not be knowledgeable about recent work in your field. Such was the case for the audience that McClintock faced at Cold Spring Harbor. In addition, the audience could include nontechnical professionals such as managers who may not have any idea about your work: its importance, the recent work of others in your field, or even the general theories upon which your work is based. Because these people often have the largest say in how your work is funded, they cannot be ignored in the presentation.

So, for a mixed audience, how do you design the presentation so that everyone is satisfied? The answer is not simple. If your goal is to satisfy the entire audience *throughout the entire presentation*, no answer exists, except perhaps to give multiple presentations to the different audiences. However, if your goal is to satisfy everyone *by the presentation's end*, then one possible answer is to speak to the different audiences at different times in the presentation.

One strategy, which is depicted in Figure 2-3, is to begin at a shallow depth that orients everyone in the room to the subject. That orientation includes showing (not just telling) the importance of the subject. Then for each division of the presentation's middle, before diving into the new topic, you begin in the shallows where everyone in the room can follow you. During the deeper dives, many members of the nontechnical and general technical audience will not be able to stay with you, but you should bring them back into the presentation with the beginning of the next topic. At the presentation's end, you should come back to the shallows and then examine the results in a way that everyone understands. With this strategy, while the nontechnical and general technical audiences may not have followed all of the theoretical derivations or the analysis of the experimental results in the middle, everyone would have learned the main points of the presentation.

A fear that many presenters have with this strategy is that they will bore the specialists with the general information. Just because you present information that an audience already understands does not necessarily mean that you bore that audience. For instance, in his lectures to freshmen physics students in the early sixties, Richard Feynman also drew a number of professors and graduate students who were interested in his presentations about subjects that they already understood. As David L. Goodstein wrote,

> But even when he thought he was explaining things lucidly to freshmen or sophomores, it was not always really they who benefited most from what he was doing. It was more often us, scientists, physicists, professors, who would be the main ben-

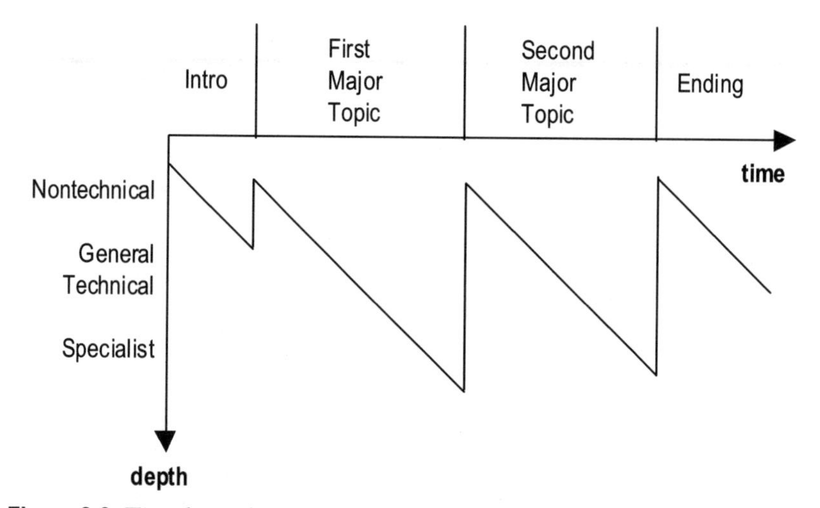

Figure 2-3. Timeline showing presenter reaching multiple audiences by beginning at surface of the topic, diving into a subject, and then surfacing to gather entire audience.

> eficiaries of his magnificent achievement, which was nothing less than to see all of physics with fresh new eyes.[12]

My colleague Dan Inman concurs with this assessment. He claims that he does not tire of listening to an explanation of something that he already knows as long as the explanation is done well.[13]

What happens when you have to speak about a subject to an audience that includes an expert who knows more than you do about one of the topics? This situation is perhaps the most intimidating. As discussed earlier, one strategy is to mention the expert by name and to admit that this person could explain the topic better than you can, but that you will try. Then you do the best that you can. By showing respect for the expert, you often recruit the expert to your side. If you say something imprecise and the expert corrects you, he or she will more than likely do so in a respectful manner.

Recognizing the Purpose

Scientific presentations have a variety of purposes. In a presentation to instruct employees of a zoo how to handle a drugged hippopotamus, the primary purpose is to inform. In a presentation to propose a purchase of a laser velocimetry system, the primary purpose is to persuade. In the opening address to a conference, the primary purpose is to inspire.

Although these mentioned presentations have clear primary purposes, most presentations carry a mixture of purposes. For instance, in a technical presentation at a conference, you not only want to inform the attendees of your work, but you also want to persuade them about your results and stimulate conversation about your subject area. Understanding the purpose of a presentation is important, because the purpose affects how you craft the speech.

Presentations to Inform. For presentations in which the primary purpose is to inform, such as instructions for handling a drugged hippopotamus, the audience typically does not doubt what you have to say. In other words, the audience does not approach this type of presentation with the same critical scrutiny as they would the presentation of new research results. Rather, the audience simply wants to learn how to perform that process. For that reason, your main objective is to deliver the information in as logical and straightforward a fashion as possible with emphasis on warnings and key steps.

For such an occasion, the adage *Tell them what you're going to tell them, tell them, and tell them what you told them* serves. The introduction places the audience in a position to comprehend the instructions, the middle simply delivers the instructions in a logical fashion, and the ending serves to increase comprehension with repetition.

Stellar examples of informative presentations occurred during the rescue of *Apollo 13*.[14] On April 13, 1970, more than halfway on its voyage to the moon, one of the oxygen tanks of *Apollo 13* exploded. Over the next three days, to bring the crew safely back to Earth, NASA had to devise and communicate a series of complex procedures to change the flight path, to adjust and readjust trajectories, and to preserve life on board the damaged ship. Further complicating matters were that the audience for those instructions was weary from lack of sleep and that all instructions had to be delivered verbally over the radio. At that time, no means existed for visual transmissions to the ship.

Presentations to Persuade. For presentations in which the primary purpose is to persuade, the challenge increases greatly. For instance, persuasion was the primary purpose of Morton Thiokol's presentation to delay the launch of the space shuttle *Challenger*.

Linus Pauling, who won a Nobel Prize in Chemistry as well as a Nobel Peace Prize, was effective in persuasive presentations. Why was that? This question is difficult to answer. Certainly, Pauling provided much logical evidence for his arguments, but as James Watson asserts about one of Pauling's presentations, Pauling also used pathos:

> Pauling's talk was made with his usual dramatic flair. The words came out as if he had been in show business all his life. A curtain kept his model hidden until near the end of his lecture, when he proudly unveiled his latest creation. Then, with his eyes twinkling, Linus explained the specific characteristics that made his model — the α-helix — uniquely beautiful.... Even if he were to say nonsense, his mesmerized students would never know because of his unquenchable self-confidence.[15]

A much different approach was taken by Maria Goeppert Mayer, who was particularly persuasive in one-on-one presentations. "Charming" is the word many people used to describe her.[16] Living in Chicago, Mayer came up with a shell model for the nucleus just as three Germans were developing a similar model. Rather than trying to beat this group by publishing first, Mayer waited and published her work at the same time. Because this shell model was such a radical departure from current thinking, she felt that two papers, rather than one, would have more influence on the scientific community. Also, rather than competing with the German group, she collaborated with one of them, Hans Jensen, on a book that explained the theory in more detail. Although she wrote most of the book, she was generous in acknowledging his contribution. What could have been a competitive situation became a fruitful collaboration. For their work, both Mayer and Jensen received the Nobel Prize in Physics.

So how is one's speech affected when the purpose is strictly to persuade? Much depends upon the initial bias of the audience toward your idea, a point that is discussed in more detail in Chapter 3. Assuming for the moment that the audience has a neutral stance to your main assertions, you have several variables to consider. For instance, not all assertions are created equal. An assertion such as, *Design A is an effective design,* is much more difficult to marshal evidence for than is the assertion, *Design C is not an effective design.* For the second presentation, all you need to do is to show that Design C does not meet one criterion for the design, while in the first presentation you have to show that Design A meets all the criteria.

Also, not all persuasive presentations call upon you to bring the audience to your position. In some presentations, the purpose is to negotiate a compromise about the situation. Maria Goeppert Mayer's situation was one in which a compromise was worked out for both parties.

Presentations to Inspire. A third purpose that arises in engineering and science presentation is to inspire an audience. Presentations that call upon you to inspire might be an opening address to a conference, an after-dinner talk, or a speech before a student organization. When the purpose of the talk is primarily to inspire, the speaker may well want to venture from the standard format of speaking for fifteen minutes with an overhead projector and stack of ten presentation slides.

An interesting example comes from a presentation delivered by Doug Henson, a manager at Sandia National Laboratories.[17] The presentation occurred at the beginning of a four-hour forum for recruiting employees to Sandia. The forum was attended by one hundred of Sandia's top management. As the opening speaker, Henson had the goal of motivating his audience behind the recruiting effort. Henson chose the following title: "Winning the War for Talent." In the beginning of his presentation, which had the difficult time slot of just after lunch,

Henson stood with his back to the audience. Then someone came out and silently outfitted him in military attire: an authentic army jacket from World War II; a leather holster with a pearl-handled revolver; a riding crop and gloves; and a helmet with insignia. At first, the audience was not quite sure what was going on. However, everyone in the room sat up and paid attention. In contrast to the underlying buzz that normally pervaded this audience, there was an intriguing silence.

After Henson was completely outfitted, a projector came on and beamed a huge U.S. flag on the wall behind him. Then Henson turned and began to speak, but not in the professional manner of a manager at a national laboratory. Rather, Henson spoke in the spirited and dramatic manner of General George S. Patton.

What Henson did was to memorize one of Patton's famous speeches. In giving the speech, though, Henson substituted Sandia's mission to recruit talented employees for Patton's mission to gain a beachhead on Italy's western coast: "You are here because you want to win. You love a winner and will not tolerate losing. I wouldn't give a hoot in hell for someone who lost and laughed, but will stake my career on someone who will fight to win." The audience listened intently to every word. At the conclusion of his speech, Henson came to attention, did a left-face, and marched off stage. Then the next scheduled speaker took the podium and began her portion of the forum, as if nothing unusual had occurred.

Henson received much good feedback for this performance. What normally would have been a sleeper presentation with eight overheads and polite applause became a provocative call for action that the audience still talked about months later. Granted, such a presentation could not be repeated to the same audience, because the power of the presentation lay in the underlying tension of the audience not knowing exactly what the speaker was doing. Another point that added to the success of this presentation was that both the chosen speech and persona were appropriate for this audience: managers at a laboratory that receives much funding from the Department of Defense.

Multipurpose Presentations. Most presentations do not have just the single purpose of informing, persuading, or inspiring. A conference presentation, for instance, certainly includes the instructional purpose of informing others about the work, but also has the purpose of persuading audiences to believe the results and the purpose of inspiring the audience to discuss the topic and contribute new ideas.

Another interesting purpose to consider is teaching a class of students. In this situation, the primary purpose is to have the students learn the material at hand, and a secondary purpose is to inspire the students to continue studying the subject after they leave the course. Given these two purposes, just telling the students the main points is not always the most effective way to teach. As a teacher, you often want the students to discover the information on their own, because by discovering the material the students are much more likely to retain the material. In other words, the students become owners of the information.

Given the wide variety of students, subjects, and methods, this book does not even attempt to discuss all the methods for teaching students. However, it is important to understand that for any given subject and audience, several different methods are effective, and at least as many methods are ineffective. Moreover, some unusual methods that would have no place in a business or conference presentation can succeed with the right teacher and audience. For instance, the great mathematics teacher Emmy Noether spoke very quickly, so quickly that the students struggled to keep up. Not only did she speak quickly, but she wiped the blackboard clean almost as soon as she had written upon it. According to one of her students, the algebraist Saunders MacLane, her method was an exercise of sorts that forced the students to think quickly, which Noether believed was necessary to become a mathematician.[18]

To introduce the first law of thermodynamics to his sophomore students, Philip Schmidt, a mechanical engineering professor at the University of Texas, uses a similar strategy to the one that Doug Henson used in the Patton presentation at Sandia. In his presentation, Schmidt dresses in the formal attire, including top hat, of Nicolas Carnot and speaks to the students as if he were Carnot himself, introducing this law for the first time. For this audience and for this occasion, the strategy succeeds.

Addressing the Occasion

In addition to considering the presentation's audience and purpose, you should think about the occasion of the presentation. The occasion is defined by several variables. One is the formality of the presentation. Is the presentation at a conference, at a business meeting, or after a dinner in a banquet hall? Each of these presentations is quite different in regard to the formality expected by the audience.

The occasion is also defined by the time limits. In some presentations, such as at conferences, the time limits are fixed because others are waiting to speak. In such situations, if you exceed the limits, you risk upsetting, even angering, your audience.

The occasion is also defined by the time at which the presentation occurs. Are you speaking in mid-morning, when people have much energy, or late in the afternoon, when people are usually tired? This variable might affect how ambitious you are, covering four main points in the mid-morning as opposed to covering just three points in the late afternoon.

Yet another defining variable is the logistics for the presentation. Is it a face-to-face meeting, as it was between the engineers and management at Morton Thiokol on the afternoon before the fateful decision to launch the space shuttle *Challenger*? Or is it a teleconference, as it was between Morton Thiokol and NASA later that evening? The logistics might affect variables such as how you design your presentation slides. In a teleconference presentation, in which you do not have the opportunity to gauge the audience's expressions and adjust your speech, you should design your presentation slides so that they stand alone.

Still other variables that define the occasion are the location for the presentation and the number of people in attendance. For instance, if the presentation is a dissertation defense, is the presentation before the dissertation committee in a small conference room? Or is the presentation in the Sorbonne before one thousand spectators, as was the case for Irène Curie in 1925?[19] The number of people in the room could affect decisions such as whether to incorporate humor. With a packed room, because the laughter of the audience appears amplified, the audience is more likely to perceive the humor as successful. If the room is half empty, though, any laughter quickly dissipates.

In summary, occasion dramatically affects the speech. If the occasion is formal and if the time short, then you probably would choose a speech that simply presents the facts and arguments. If the occasion is informal or if there is time to diverge from a "just-the-facts" style, you might work in different flavors to the speech: anecdotes, examples, stories, humor, and personal connections.

Critical Error 2
Drawing Words from the Wrong Well

It is hard to overestimate the dismay and resentment of an audience that has to put up with a paper read hurriedly in an even monotone.[1]

—P. B. Medawar

Toward the end of World War II, Niels Bohr set up a meeting with Winston Churchill to warn him about an atomic arms race that Bohr correctly predicted would occur after the war. Bohr wanted all countries to establish guidelines to contain these weapons. A few months earlier, Churchill had diminished hopes for such guidelines by signing away British rights to nuclear development. Because Bohr, who had recently fled Denmark, did not speak English very well, he decided to write out the presentation in his best English and have a friend, R.V. Jones, go over the draft and polish the language. For three days, the two men worked on this presentation, and when Bohr was pleased with the product, he memorized it. The day of the meeting arrived, and Bohr was brought to Churchill by an aide who was sympathetic to Bohr's position. Unfortunately, as soon as Bohr and the aide met Churchill, Churchill put both the aide and Bohr on the defensive by claiming that the meeting was nothing more than a reproach for England's signing away of the rights to nuclear development. Bohr tried to improvise, but according to R.V. Jones, "no doubt suffered from his usual anxiety to be precise."[2] Within twenty minutes, Churchill lost patience and had Bohr ushered out of the office.

How should scientists and engineers deliver their words in a scientific presentation? Should they read those words, memorize those words, or speak from notes or presentation slides? Or should they just speak off the cuff; in other words, should they not worry about what they have to say until they are standing before the audience? Before deciding upon an answer, you should consider the advantages and disadvantages of each source of words, as listed in Table 2-1, and the occasions to use each, as shown in Table 2-2.

Speaking from Points

By far, the most common and accepted way to make a scientific presentation is to speak from points that you have memorized or have placed onto slides or written down as notes. P.B. Medawar strongly recommended this strategy,[3] as did Michael Faraday.[4] Another proponent of this strategy was Richard Feynman. For instance, for his famous set of lectures on freshman physics, Feynman brought to each class only one sheet of notes. Einstein used this strategy for his lectures as well, bringing to class only one note card.[5] The advantages of this strategy are numerous, perhaps the most important being the effect upon the audience in regard to the audience's assessment of the speaker. Because the presenter is producing most of the words from within himself or herself, the audience perceives that the speaker owns this information, as opposed to having been given this information.

Rather than speaking from a page of notes or a set of presentation slides, some speakers simply memorize the points that they are to make. Boltzmann apparently used that method, and not just for a few lectures, but for a series of lectures that spanned four years and included such varied topics as classical mechanics, hydrodynamics, elasticity theory, electrodynamics, and the kinetic theory of gases.[6]

Table 2-1. Advantages and disadvantages of different sources for speech.

Sources	**Advantages**	**Disadvantages**
Speaking from points	Credibility earned Ease of adjusting speech Eye contact Natural pace	Wording not exact Long preparation time
Memorizing	Precision Smooth delivery Credibility earned Eye contact	Potential for disaster Unnatural pace Inability to adjust speech Long preparation time
Reading	Precision Smooth delivery	Credibility undercut Lack of eye contact Unnatural pace Inability to adjust speech Long preparation time
Speaking off the cuff	No preparation time Eye contact Natural pace	Potential for disaster Difficulty in organizing Lack of visual aids

Table 2-2. Situations appropriate for each source of speech.

Sources	**Situation**
Speaking from points	Conference presentation Presentation at business meeting University lecture
Memorizing	First few words of presentation Short introduction of a speaker
Reading	Press conference Quotation within a presentation Complex wording within presentation
Speaking off the cuff	Answering a question Asking a question

Another advantage of speaking from points is that because the speaker must find the words from within, he or she ends up working through the subject at a pace that is much closer to the way that the audience understands the material. In other words, when the speaker comes upon a difficult point, the speaker naturally slows to explain that point because the words do not come as easily. Paralleling that decrease of the speaker's pace is the decrease of the audience's comprehension rate. The more difficult the idea, the more time the audience needs to understand that idea. Similarly, when the speaker covers material that is relatively easy, the words come more easily, but that is fine for the audience because the understanding comes more easily.

Yet another advantage of this strategy is that the speaker has ample opportunity to make eye contact with the audience. Because the lion's share of the wording comes from within, the speaker can keep his or her eyes trained upon the audience. That opportunity allows the speaker to read the audience and to adapt the presentation to their understanding or lack of understanding.

A final advantage is that because the words are not set in stone, the speaker can change the presentation to accommodate the audience. Should the speaker perceive that the audience does not understand something or that the audience is bored and wants the presentation to move more quickly, the speaker can make the desired adjustments.

The main disadvantage with this strategy is that because the words are not set in stone, the speaker runs the risk of not having the exact words during the presentation. The speaker might become stuck as he or she gropes for the right word. To counter this disadvantage, the speaker should, as Medawar suggested,[7] practice the presentation repeatedly until the speaker is sure that the words will come. Another counter to this disadvantage is that an audience for a scientific presentation does not expect the words to flow as from an actor in a dramatic performance. If the speaker in a scientific presentation must pause to come up with the right word, the audience does not judge the speaker harshly. In fact, such pauses if properly spaced can emphasize key points. Also, if the speaker desires exact wording, say for a difficult concept or for the incorporation of a law or statute, the speaker can include that exact working on the slides or in the notes.

Another disadvantage of speaking from points is that the preparation time is generally higher than for simply reading. The reason is that for the speaker to gain confidence that the words will come, the speaker has to practice the presentation several times.

Memorizing a Speech

One advantage of having memorized a speech is that the speaker can deliver the words in a dramatic fashion, as an experienced actor does in a play. Another advantage is that because the words come from within, the speaker can maintain constant eye contact with the audience. Yet a third advantage is that because the speaker chooses the words beforehand, the speaker has control over the exact wording, as long as the speaker's memory does not fail.

A major disadvantage of memorizing a speech is that for most of us, memorizing a speech takes too much time. In a presentation, the typical person says more than one hundred words per minute. For that reason, a fifteen-minute presentation then calls for memorizing more than fifteen hundred words. That is quite a task! Screen actors in supporting roles have won academy awards for saying fewer words. Given the frequency with which scientists and engineers have to make presentations, most scientists and engineers simply do not have the time to memorize their presentations.

Another disadvantage of memorizing a presentation is that memorization does not leave much opportunity for changing the presentation in midstream, which is one of the reasons that scientists and engineers make presentations about their work, as opposed to just documenting their work in writing. In fact, Bohr's failed presentation to Churchill suffered for this very reason.

Yet another disadvantage of memorization is that the pace of our recall of words from memory does not necessarily reflect the pace at which the audience understands those words. Stated another way, our memory might recall a sentence more quickly than the audience can understand that sentence.

Given these disadvantages, you might think that memorization has no place in scientific presentations. That is not true. When you have only a few words to say before an audience, such as the introduction of a colleague, memorization might be the best approach. Also, you might memorize the first couple of lines of a difficult or important presentation just so that you create a good first impression with the audience and so that the words begin to flow as you speak from your slides or notes. For many people who speak from points, the first couple of sentences are the most difficult. Much of that difficulty arises from the nervousness that speakers often feel before a presentation. Having the first couple of lines memorized allows you to get started and to get to what Feynman refers to as that miraculous moment when you concentrate on the science and are "completely immune to being nervous."[8]

Reading a Speech

The principal advantage of reading a speech is that you say the exact words that you intend to say. As you can imagine, given the disdain that so many engineers and scientists such as Medawar[9] and Faraday[10] have for speeches that are read, the disadvantages are numerous.

As with a speech that someone gives from memory, a speech that someone reads often is at too fast a pace for the audience to understand. Complex ideas that should be presented slowly are often rattled off. Moreover, when someone reads a speech, that person's eye contact is on the page and not on the audience. The lack of eye contact prevents the speaker from assessing the reactions of the audience. The lack of eye contact also prevents the audience from assessing the intentions of the speaker. The audience gathers much from the eyes of the speaker in terms of emphasis. When the speaker's eyes are on the page, the audience cannot read those eyes.

Another disadvantage is that when someone reads a speech, the audience wonders whether the speaker actually knows the subject or is repeating what others have gathered. Granted, some disciplines such as literary criticism have a tradition of reading papers at conferences. For those disciplines, a read speech does not cast shadows on the credibility of the speaker in the same way that a speech read in the sciences or in engineering does. Yet another disadvantage of reading a speech is that changing the presentation is more difficult to do. Because the speech is already ordered on the page, rearranging that speech poses problems.

Although reading a speech has many disadvantages in a conference presentation, business meeting, or university lecture, its one main advantage (precision) might cause you to choose this source in a press conference about a controversial issue. In such a situation, where the audience scrutinizes every word or phrase, the precision that a read speech offers can outweigh the disadvantages. Reading would prevent slip-ups such as the one about the United Negro College Fund that continues to haunt former Vice-President Dan Quayle. Instead of repeating the fund's slogan, "A mind is a terrible thing to waste," Quayle inadvertently said, "What a waste it is to lose one's mind or not to have a mind is being very wasteful."[11] In searching for the right words, everyone on occasion makes mistakes. Quayle's mistake was embarrassing. What made the mistake inexcusable, though, was that he knew beforehand that his critics from the *New York Times* and other publications would be present and monitoring his every word.[12] When the exact wording of scientists and engineers is under such scrutiny, then reading a prepared statement makes sense.

Speaking off the Cuff

The principal advantage of speaking off the cuff or extemporizing is that you do not have to spend any time in preparation. If a second advantage exists, it is that the speaking pace of the presenter parallels the comprehension pace of the audience.

Given the cost, though, in assembling a professional audience for a presentation, such a strategy for an entire presentation is unsound. For a complex subject, the likelihood is low that an extemporaneous speaker would come up with an efficient and effective structure that emphasizes the most important points and that makes smooth transitions between those points. Moreover, the potential for disaster—the speaker becoming lost—is high. This potential for disaster might also be a deep source of nervousness for the speaker. After all, the best countermeasure against nervousness is preparation. Also, with extemporizing, the likelihood is high that the speaker will lose the audience, should the presentation go into any depth. Yet another disadvantage is that little chance exists for visual aids, other than a writing board.

Although such a strategy is discouraged for a conference presentation, lecture, or business meeting, practicing short extemporaneous talks is time well spent, because often in conference presentations, lectures, and business meetings, one is forced to extemporize during question periods. The more practice that a presenter has at speaking extemporaneously, the more confidence that presenter is likely to exhibit during question periods.

When asked a question, it is important to pause and think before answering. Such a pause not only allows you to consider what you will say, but also provides emphasis to the first sentence of your answer. The audience is patient with a speaker who silently thinks about the question for a moment, much more patient than if the speaker fills the silence with empty chatter or a filler phrase such as *uh, um*, or *you know*.

How do you eliminate filler phrases from your speech? The process generally takes several days, with the first step being to learn what filler phrases you say. That step you can accomplish by having a colleague critique a presentation of yours. Once you have discovered what your filler phrases are, your subconscious will work to eliminate them from your speech. Your subconscious is powerful. Do not underestimate its abilities. When you notice yourself saying one of your filler phrases, you are well on your way to eliminating that phrase from your speech. Not surprisingly, you are much more likely to say filler phrases when you are tired, which is reason enough to get a good night's sleep before an important presentation.

Chapter 3

Structure: The Strategy You Choose

Whereas Einstein tried to grasp a hidden essence by disregarding anything he thought irrelevant, Bohr insisted that nothing be left out.[1]

—Edward MacKinnon

The success of a presentation hinges on its structure. In a presentation, structure comprises the organization of the major points, the transitions between those points, the depth that the presenter achieves, and the emphasis of details. Although many aspects of the structure of a presentation are similar to that of a document,[2] three aspects are not. One difference is the necessity of the speaker to begin at a depth that orients the entire audience. In a document, if the author mistakenly begins at a level of understanding that is too deep for the readers, the readers have the opportunity to look up background information and still comprehend the document. In a presentation, though, this opportunity does not exist.

Another structural difference between presentations and documents is the importance of the speaker mapping the presentation for the audience. Granted, mapping is important in a document, but in a presentation, mapping is crucial. While readers of a document have the chance to scan ahead to see the upcoming sections and to see how long the sections are, the audience of a presentation does not.

Yet a third structural difference between presentations and documents is the importance of the speaker signaling transitions between major parts of a presentation. In a document, the audience can anticipate transitions to new topics from the headings, subheadings, and paragraph breaks. In a presentation, though, the audience depends entirely upon the presenter for those transitions. Presenters can signal topic transitions with phrases in speech, changes in slides, or pauses in the delivery.

This chapter focuses on these three differences and points out two major pitfalls to avoid: leaving the audience behind at the beginning and losing the audience in the middle.

Organization of Presentations

One way to look at the organization of a presentation is to examine its beginning, middle, and ending. In this organization, which is represented by Figure 3-1, the beginning shows the big picture of the presentation and then focuses everyone's attention to the particular topic.

The middle then discusses the topic in a logical fashion. If the topic is a process, such as the evolution of a Hawaiian volcano (represented in Figure 3-2),[3] a logical strategy would be chronological. If the topic is an event, such as the eruption of Mount St. Helens in Figure 3-3,[4] a logical strategy would be spatial: the flows of lava down the mountain and the plume of smoke and ash into the atmosphere. If the topic is a system, such as the solar power plant in Figure 3-4,[5] then a logical strategy would be the flow of energy through the system.

If the topic can be classified into parts, a logical strategy often is a grouping into parts that are parallel. An

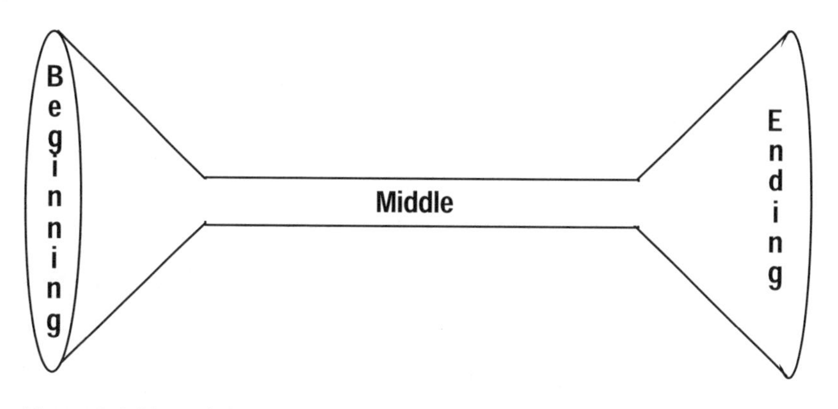

Figure 3-1. Visual depiction of the organization of a scientific presentation. The speaker begins with the big picture, focuses on the work in the middle, and comes back out to the big picture in the ending. In essence, the ending discusses the repercussions of the work on the big picture, which was introduced in the presentation's beginning.

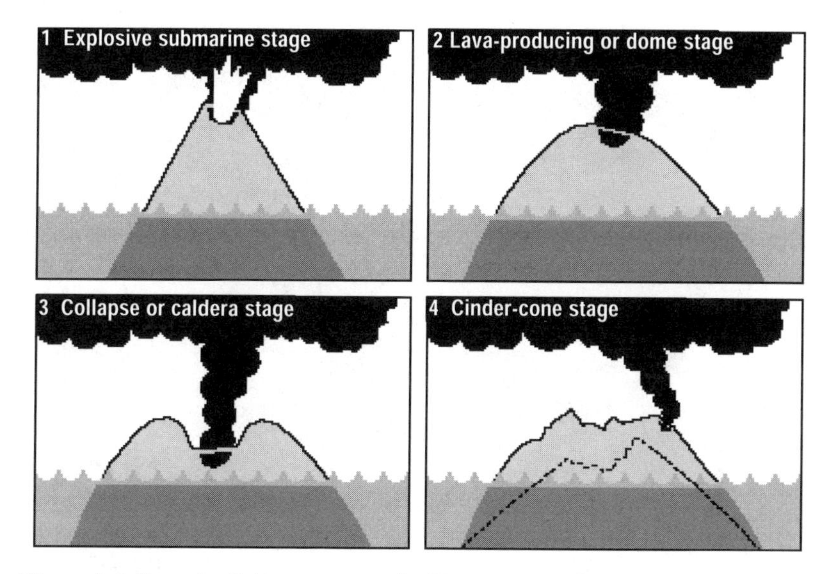

Figure 3-2. Four building stages of a Hawaiian volcano.[3] In presenting these stages, a logical strategy would be chronological (from the first stage to the fourth).

Figure 3-3. Eruption of Mount St. Helens (courtesy of United States Geological Survey).[4] One way to describe this event would be spatially, tracking both the flows of lava down the mountain and the plume of smoke and ash into the atmosphere.

Figure 3-4. Solar One Power Plant located near Barstow, California.[5] In explaining the operation of this plant in a presentation, a logical strategy would be to follow the flow of energy through the system: Sunlight strikes the mirrors and reflects onto the receiver mounted on top of the tower. This radiant energy is converted to heat in a transfer fluid that flows to a turbine to produce electrical energy.

example, shown in Figure 3-5, is the classification of methods for reducing the emissions of sulfur dioxide from coal power plants.[6] One way to classify those methods would be into precombustion methods, combustion methods, and postcombustion methods. Note that for each category, the presenter might have subcategories. For instance, two common postcombustion methods are adsorption and absorption.

In the ending to a presentation, the speaker analyzes the work from an overall perspective. This analysis usually contains a summary of the most important details of the work.[7] The analysis often also contains information that provides closure to the presentation: a set of recommendations for the work, a list of questions about the work that still need to be resolved, or an examination of how the work presented in the presentation's middle affects the big picture presented in the beginning.

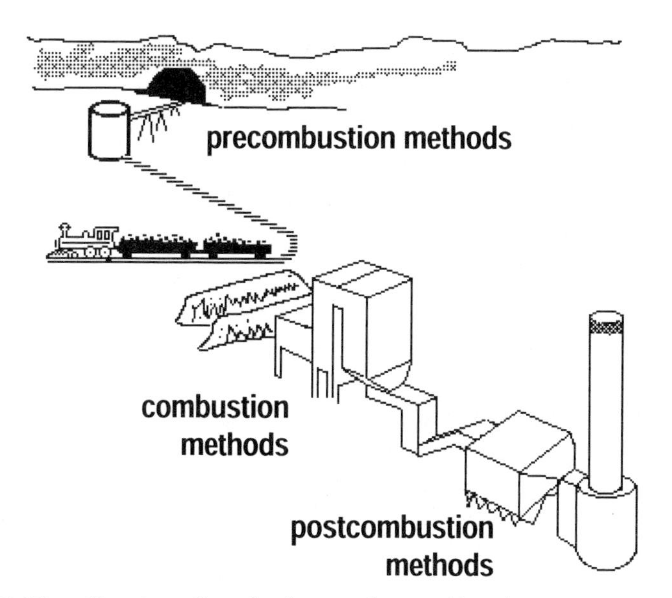

Figure 3-5. Classification of methods to reduce sulfur dioxide emissions from coal power plants.[6] A precombustion method is coal cleaning, a combustion method is an atmospheric fluidized bed, and a postcombustion method is absorption.

The failure of a speaker to organize a presentation logically sounds a death knell on the likelihood that a shrewd audience will return to listen to that speaker. For instance, lack of organization was the reason given by Linus Pauling for continually skipping the lectures of the physicist Robert Millikan.[8]

Transitions in a Presentation

Presentations have several key places, shown with arrows in Figure 3-6, where speakers have to make key transitions. The first major transition occurs between the introduction and the middle of the presentation. Usually, the middle is broken into two, three, or four parallel sections. If, for example, the presentation's middle has three sections, transitions would then occur between the first and second sections and between the second and third sections. A final transition would occur between the end of the middle and the conclusion. This transition is particularly important for the audience to recognize, because

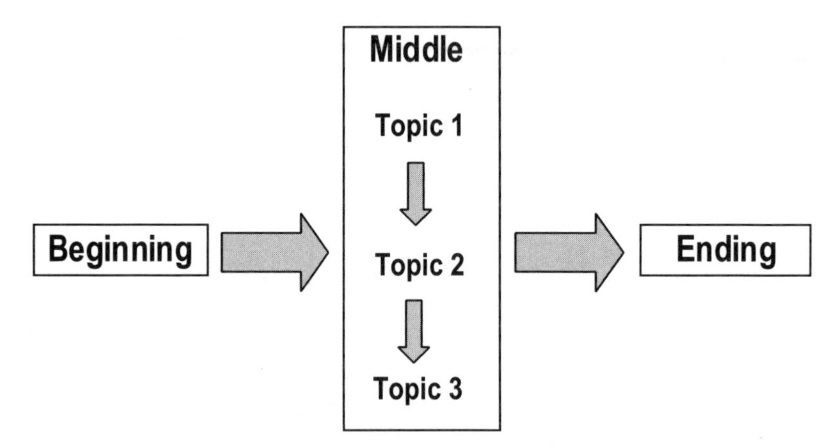

Figure 3-6. Key transition points in a presentation that has three main topics in the middle. Major transitions occur between the beginning and middle and between the middle and ending. Other transitions occur between the main topics of the middle.

if the audience realizes that the ending is at hand, they will sit up and pay more attention. After all, only a minute or two more of concentration is expected. In a way, the energy level of the audience picks up at this point much as a stable mare's pace picks up once it returns to within sight of the stables.

Making a transition becomes an even greater challenge when the change in topic is accompanied by a change in the speaker. During a group presentation, each new speaker requires the audience to adjust to a new voice, stage presence, and set of movements. While occasional changes in speaker can serve a long presentation by providing variety, too many changes in the speaker in a short presentation can cause confusion.

Depth of Presentations

In my own surveys of scientists and engineers, the most commonly asked question about the structure of presentations is, How much depth should the speaker go into? Although the simple answer to the question is, whatever depth the audience needs or desires, determining this depth is not easy. Moreover, in presentations to multiple audiences, the difficulty in determining this depth increases severalfold.

As discussed in Chapter 2, meeting the expectations of your audience in a presentation requires imagination and sensitivity. You have to assess what the audience already knows about the subject and begin your discussion at that level. Also, you have to account for how much the audience wants to learn, or needs to learn, about the subject. Because the audience will not be interested in every bolt that you turned, you have to pick and choose details. This picking and choosing is difficult, which is why that question arises so often in my surveys.

Depth is interwoven with scope, which consists of

the boundaries of the presentations. In other words, how broad a topic does the speaker address? In many presentations, such as a progress report on a project, the presentation's scope is already determined. In other presentations, such as a research seminar, the speaker determines the scope. In general, the wider the scope, the more difficult it is to satisfy the audience. The relationship of scope and depth can be seen in the dimensions of vessels of equal volumes, as depicted in Figure 3-7. For a presentation with a fixed time (a vessel with a fixed volume), the speaker can convey only so many details. For that reason, the wider the scope (as in the top vessel), the less the depth that the speaker can achieve and the more the potential questions that loom for the audience. Likewise, the narrower the scope (as in the bottom vessel), the more the depth that the speaker can achieve and the more likely that the speaker will answer the questions about that topic.

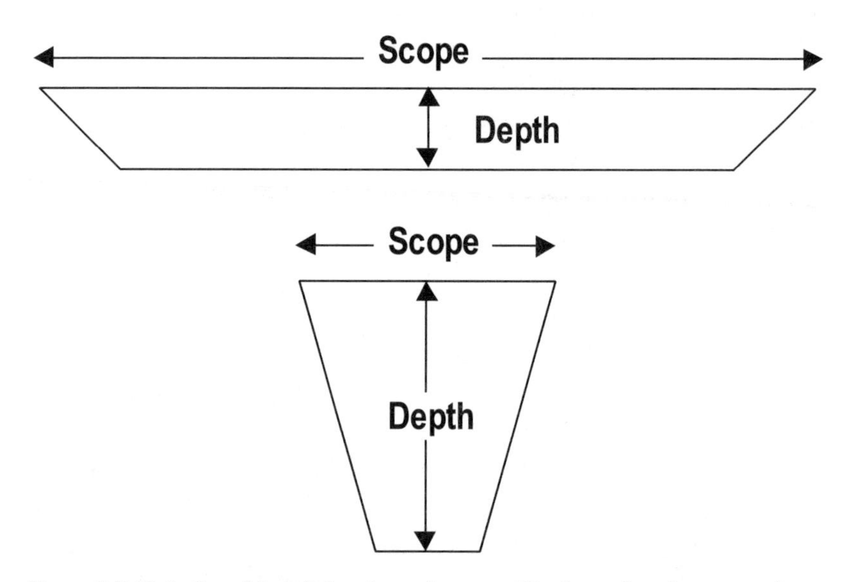

Figure 3-7. Relationship of depth and scope. The broader the scope (top image), the less the depth that the speaker can achieve and the more difficult it becomes to satisfy the audience. Likewise, the narrower the scope (bottom image), the more the depth that the speaker can achieve.

Just because giving a "broad-scope" presentation is difficult does not mean that one should avoid giving such presentations. Rather, giving a broad-scope presentation means that the challenge is greater and the speaker has to think long and hard about the presentation's structure.

Recently, I attended a NASA presentation entitled "Eight Technology Innovations for Space Applications." Although the speaker was knowledgeable, engaging, and deft at explaining concepts, the presentation was unsatisfying. Granted, the first twenty minutes were interesting. During this time, the speaker discussed high-temperature superconductors, which was the first innovation on the list. Unfortunately, in the remaining forty minutes, the speaker tried in vain to cover the other seven innovations. Throughout this portion, the speaker furiously flipped through a huge stack of overhead slides, selecting one every so often to project, showing it for less than 15 seconds, and then apologizing for how little time he had.

One problem with this NASA presentation was that all the slides were much too detailed for the depth that the speaker was forced to adopt. The effect of showing such detailed slides in such a cursory manner was that the audience had many unanswered questions about each one. In the end, the audience became frustrated because the speech was so shallow and yet the slides were so deep. If the speaker had limited himself to only two or three innovations that have space applications, then the presentation could have succeeded. Another possible successful strategy would have been to group the eight innovations into two or three memorable categories and then create more general slides for those categories. This second strategy would have been more challenging, but given how dynamic and knowledgeable the speaker was, it could have succeeded.

Emphasis in Presentations

The emphasis of details in a scientific presentation is as important as the organization of details. On average, people remember only about ten percent of what they hear.[9] For that reason, although a presentation might be well organized, the presentation could fail without proper emphasis. In such a case, the audience could walk out of the room remembering only the ten percent of details in the presentation that were least important.

Much about the way you emphasize details in a presentation is similar to the way you emphasize details in an article or report.[10] Repetition, illustration, and placement play important roles in both situations. One key place for emphasis in a presentation is the beginning, when the audience is the least tired and the listening abilities are the sharpest. For that reason, you want to use the beginning of a presentation to say something important: to define the scope of your presentation, to show the audience the importance of the work, and to map in a memorable way the path of the presentation.

Although the middle of a presentation is not a place in which audiences typically listen more closely, you can help maintain a higher level of retention for the audience if the audience sees the logical path that you have chosen to explain your work. As mentioned in the section on organization, that path might be chronological or spatial; it might follow the flow of a variable through a system; or it might break down the presentation topic into parallel divisions.

Whatever path you choose, the number of divisions in the path should not be large. People remember groups of twos, threes, and fours. Groups larger than four tax the listener. I cannot count the number of times that I have seen a presenter say that he or she will discuss six, seven, eight, or even nine main points. In such presentations, the audience members often take deep sighs and glance at their watches. For the audience, the chances for retention have dramatically dropped.

The ending is also a wonderful opportunity for emphasis, especially if the audience knows that the ending is upon them. Why is that? As mentioned in the section on transitions, if the audience knows that the ending is near, they will sit up and concentrate, even if they have not understood everything up to that point. You can observe this phenomenon at church, especially in Protestant churches in the South in which the sermons go for thirty minutes or longer. In those sermons, the preacher usually gives the congregation a clue that the end of the sermon is close at hand: "As is sung in the hymn of invitation, number 343, 'On a Hill Far Away,'" and so forth. At that moment, the congregation realizes that the sermon will end in a couple of minutes. In many a sermon at that point I have distinctly heard the creaking of pews as everyone sat up.

Another missed opportunity for emphasis in the ending occurs when engineers and scientists fail to leave their conclusion slide projected during the presentation's question period. The conclusion slide is the most important slide of the presentation, because it summarizes the presentation's main results. For that reason, a presenter should want to show this slide as long as possible. However, many presenters show this slide for only a minute, or less in some cases. What, then, do the presenters project during the three to five minutes of the question period? Inexplicably, many presenters project a question mark or other worthless projection. Because a question period has dead time whenever a tangential question is posed, the audience has time to focus attention back onto the screen. By continuing to show your conclusion slide during the question period, you increase the chances that the audience will retain your presentation's most important results.

Critical Error 3
Leaving the Audience at the Dock

> *I think from all I hear [that] I was a very difficult lecturer. I started as a lecturer who made things very difficult. I had some help; I remember [Wolfgang] Pauli's advice, almost certainly in '28. He said, 'When you want to give a seminar or lecture, decide what it is you want to talk about and then find some agreeable subject of contemplation not remotely related to your lecture and then interrupt that from time to time to say a few words.' So you can see how bad it must have been.*[1]
>
> —J. Robert Oppenheimer

How many times have you seen a presenter project a title slide and then quickly remove it before you had the chance to fathom what the presentation was really about? Or how many times have you had a presenter overwhelm you with an outline list of presentation topics, more than half of which you were unable to remember not one minute after the mapping slide had been removed? Such are the trademarks of presentations in which the presenter leaves the audience behind at the beginning.

Becoming lost at the beginning of a presentation is frustrating for audiences. When the presentation format does not allow for questions until the end or when the size of the audience inhibits the audience from asking for clarification, becoming lost at the beginning often means that the time spent at that presentation is wasted. For that reason, presenters should make it a goal to make the beginning as clear as possible.

One reason that beginnings to scientific presentations often fail is that the speaker has not anticipated the initial questions of the audience. What is the presentation about? Why is the presentation important? What knowledge is needed to understand the presentation? How will the presentation be arranged? Another reason that beginnings to scientific presentations often fail is that the speaker has not anticipated the biases of the audience. Although the speaker might target the level of the audience in his or her arguments, if the speaker does not account for the biases of the audience, he or she may adopt an inappropriate strategy or not marshal enough evidence to support the presentation's assertions.

Anticipating the Audience's Initial Questions

While most in attendance at a scientific presentation do not expect to understand everything that the presenter puts forward during the presentation, they hope to understand at least something for the time invested. Unfortunately, many scientific presenters begin as if the audience has had nothing better to do for the previous week than to read every paper that the presenter has written. Why would a presenter make such an assumption? I suspect that in many cases the answer to this question is fear—fear of being considered simplistic. It takes courage to orient the audience to what you have done, because once you have, the audience is in a position to critique your efforts.

So what makes for a strong beginning of a scientific presentation? Imagine yourself in the audience. A few hours or perhaps days earlier, the presenter's title and summary interested you, but now the specific details seem cloudy. From the time you first read the title and abstract and decided to attend the presentation, events have occurred; for instance, at a conference, other presentations have taken place. As the presenter moves to the front of the room, a hush falls over the crowd, and the following questions come to you:

(1) What exactly is the subject?
(2) Why is this subject important?
(3) What background is needed to understand the subject?
(4) In what order will the subject be presented?

A strong beginning to a scientific presentation answers these questions.

Now, in some presentations, the speaker does not have to explicitly address all of these questions at the beginning. For instance, the audiences of some presentations might already know the answers to one or even two of these questions. The point is that by the time the presentation's introduction is over, none of these questions should be hovering over the audience.

What about the situation in which the audience is diverse? Often, with a diverse audience, some in the audience will not need to have question two or three answered. Despite that, others will. More than any part of the presentation, the beginning should target the widest possible audience. Although many in that audience will not follow you through every detail in the presentation's middle, if you have a strong beginning, everyone in the audience should be able to go back to his or her colleagues and summarize in a general way what you have done and why the work was important.

How much time should you spend on the introduction? Much here depends upon how much time you have for the entire talk. In a sixty-minute talk, an audience will accept ten minutes on the introduction. In a fifteen-minute conference presentation, though, you should limit yourself to no more than five minutes. Spending too much time on the introduction makes the audience impatient. According to my colleague Dan Inman, when a speaker tells a long story in the introduction of a conference presentation, many people assume that the speaker does not have much to report.[2]

More important than time, though, is the understanding of the audience. According to the Polish physicist Leopold Infeld, Einstein was adept at introductions. Einstein had a "calmness" that contrasted sharply with the "restlessness" that many presenters showed. These restless presenters mistakenly assumed that the audience was equally familiar with the subject matter and proceeded quickly into the details of the talk.[3] Einstein, on the other hand, patiently prepared the audience for the problems about which he would speak.

What Exactly Is the Subject? The first question that an audience has about a technical presentation is, "What is this presentation about?" The answer to this question directs the audience to what they should learn from the presentation. Many speakers, unfortunately, do not give satisfactory answers to this question in their presentations. Perhaps, many of these speakers assume that the audience already knows what the presentation is about from the posted title and abstract.

Such an assumption is dangerous. Even if the audience has already read a title and abstract of the presentation, they very likely have done so hours or even days before. Also, because most people remember only about twenty percent of what they read, a review is usually welcomed. Moreover, because many things probably have happened between their reading of the title and abstract and the beginning of the presentation—particularly at a conference—the title and abstract are not in the forefront of a listener's mind. Finally, the speaker's abilities to emphasize details through the voice's loudness and cadence provide a useful perspective of those details for the audience.

A common mistake of speakers is to go over the answer to this first question too quickly. How many times have you been to a presentation in which the speaker places the title slide up on the screen and discusses it for only fifteen or twenty seconds? That is too short. When you first begin to speak, the audience has to adjust to your delivery style: your voice, your movements, facial expressions. Such adjustments take a while, and during that time the speaker has to be careful not to overwhelm the audience with too much information.

As regards title slides, such as the one in Figure 3-8, some information is mandatory: title, name of speaker, and affiliation of speaker. Other information should be included if possible or applicable: key image to orient the audience, date, institution logo, and name of sponsor for the work. The key image is one that many presenters neglect, but it is an important opportunity to orient the audience. Having the image allows you to speak longer about the subject without feeling compelled to change slides. For a slide to be effective, it should re-

Figure 3-8. Sample title slide.[4] Notice that this slide includes not only the presentation's title and information about the speaker, but also a key image from which the speaker can discuss background.

main on the screen for at least sixty seconds. That way, the audience has the opportunity to take in the information and still listen to what you have to say. When a photograph does not fit onto the slide, the speaker should consider a watermark background, such as the row of turbine vanes shown in Figure 3-9. Although the textures and colors of a watermark are not nearly as clear as in a photograph, a well-chosen watermark still allows the audience to view a key image in the presentation. Be careful about watermarks, though. If the audience does not recognize the image and if the speaker does not remember to identify the image, then the audience can become distracted to the point of missing key details in the speech.

Why Is the Subject Important? Listening to a presentation is difficult work, so difficult that audiences will give up

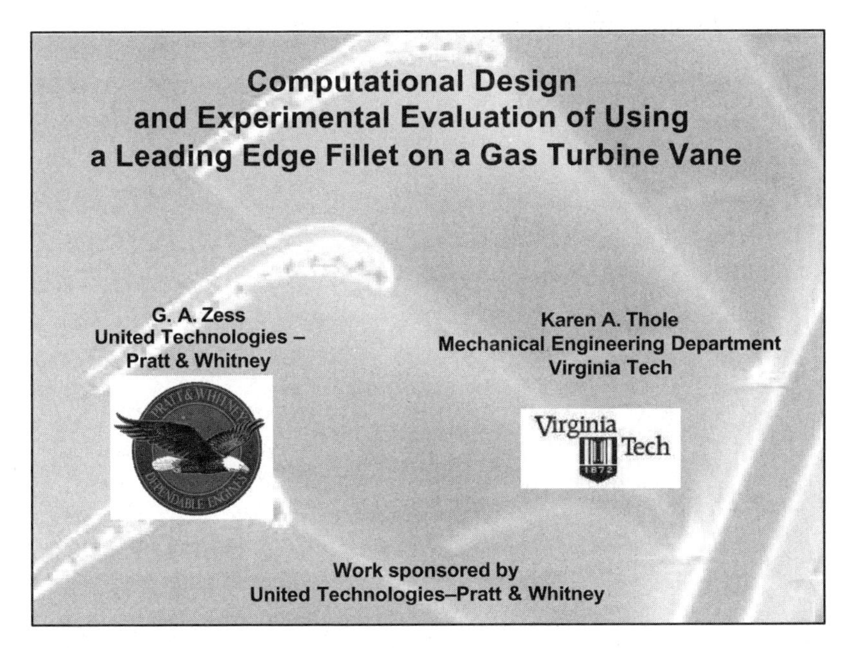

Figure 3-9. Sample title slide.[5] Notice that this slide includes not only the presentation's title and information about the speaker, but also the key image of a row of turbine vanes as a watermark.

concentrating if they do not have sufficient reason to do so. Given that, you should not move into the middle of your presentation unless you are sure that your audience understands the importance of your presentation. Often, the importance relates to money, safety, health, or the environment. For example, in a presentation comparing techniques to reduce sulfur dioxide emissions from power plants that burn coal, you could establish the work's importance by stating how many tons of sulfur dioxide are emitted each year by these power plants. Then you could discuss the contribution of sulfur dioxide to acid rain and the deleterious effects of acid rain on lakes, streams, foliage, and statues. Such connections would take but thirty or forty seconds to make, but would mean much to an audience when energy is required of them to stay with you in the middle.

In other situations, the issue is not so much the importance of the subject, but curiosity about the subject. For example, consider a presentation proposing research to study Jupiter's two largest moons: Ganymede and Callisto. Such research has no direct benefit in terms of money, safety, health, or the environment. Rather, the main reason for the researcher to study these moons is curiosity. To give the audience the motivation to stay with the presentation for its duration, the speaker should instill in the audience the same curiosity that he or she has for the work. For instance, the speaker could explain that Ganymede and Callisto are of about the same size and density. The difference between the moons is their color: Callisto is dark all over, while Ganymede has dark patches separated by light streaks.[6] Then the speaker could raise the question of why that difference in color exists, thereby making the audience curious.

Often, stating the importance of a subject involves grounding the problem in a specific example. Such a grounding helps many in the audience to stay with you through the abstract or mathematical parts of your presentation. Using an example in this way was a favorite technique of Richard Feynman.[7]

What Background Is Needed to Understand the Subject? An unspoken fear that many audience members have about attending a scientific presentation is that they will not understand the subject. All too often, audiences find themselves sitting in scientific presentations and having no real idea what is being discussed. Such a situation is frustrating, particularly when the format does not allow the audience to ask a question until the presentation's end. Even if audience can ask questions, many in the audience will not do so for fear that the question would distract the rest of the audience or for fear that the rest of the audience might think them ignorant.

Because different types of audiences often attend scientific presentations, a speaker should be sensitive to the background information that audiences need to understand the presentation. How do you know what background information to provide? This question is not easy to answer. Sometimes, the time limit of the presentation is such that you have few options. In such cases, it is often important to state up front what you are assuming that the audience knows. That way, those who do not have that information can set reasonable expectations for what they will comprehend. By knowing that they will not understand all of the presentation, the audience can prepare itself to receive a reduced amount. In a way, that knowledge allows the audience members to relax and perhaps to understand more than they would if they tried to follow every step.

Note that a speaker does not have to give the audience all of the necessary background in the introduction of the presentation. Another possibility is to provide background as the audience needs it during the presentation. In such a case, the introduction is still a wonderful opportunity to clue in the audience to what that background information will be given so as to allay any fears by the audience that they will not be able to understand the presentation. Also, some background details, such as the major assumptions of the work, are better placed up front. At times, that kind of background can be so long that it appears to the audience as if it is a separate section of the middle. It is acceptable to label a background topic as a separate section as long as the audience can see the relationship of that section to the remainder of the talk.

In What Order Will the Subject Be Presented? The last of these introductory questions—how the subject will be presented—is more important in a presentation than in a document. Why? Unlike a document, in which the readers can glance ahead to see the headings and subheadings and therefore see what information will occur, the listeners to a presentation have no idea where the presentation is going unless the presenter tells them. In answering this question of how the details will be presented, the presenter reveals in essence the organization of the presentation. When the presenter clearly and memorably maps the organization, such as that depicted in Figure 3-10, the audience has a good idea at any point in the presentation about how much has been covered and how much further the presenter has to go. That knowledge is important, because listeners have to pace themselves. Listening is hard work, and asking someone to listen, especially to a scientific presentation, without giving a clue as to the path of that presentation, is similar to taking that person on a hike without naming the destination. Because the person does not know how far he or she is going, the person quickly tires.

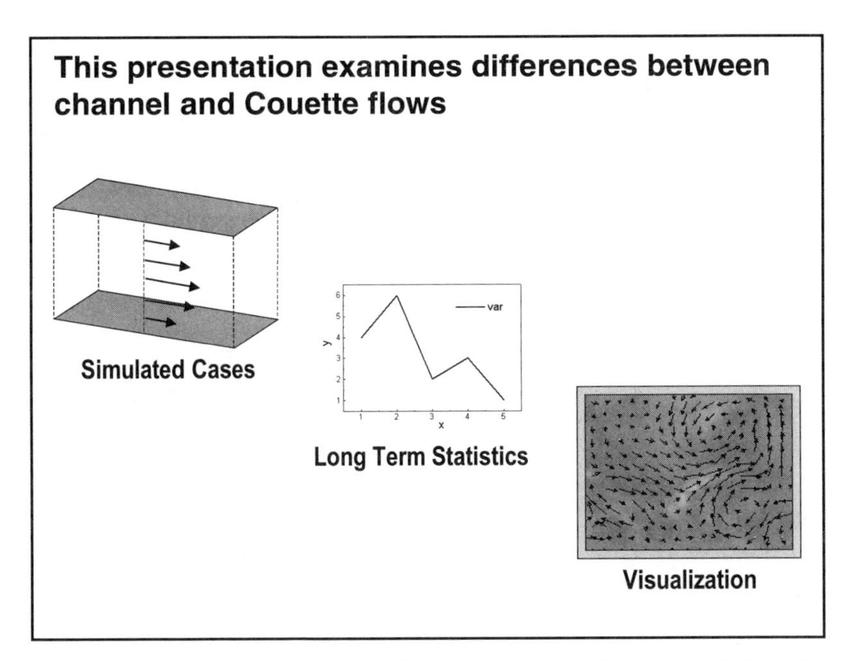

Figure 3-10. Example mapping slide.[8] One strength of this slide is its use of images to make the mapping memorable. These images are repeated in the corresponding divisions of the presentation. Another strength is that this slide dispenses with unneeded listings such as "Introduction" and "Conclusion" (every presentation has those).

Anticipating the Audience's Bias

In cases where you have to persuade an audience, an important question to ask is, What will be the initial response of the audience toward the results? The answer to this question can significantly affect both the strategy of the presentation and the amount of evidence needed to support the presentation's assertions.

The legend goes that in the 1980s a committee of US scientists was assigned to determine which areas of the country would be finalists for the location of a nuclear waste repository. Most of these places under consideration were rural locations. After carefully considering the local geography and other criteria, the committee made its selections. Before these selections were to be made public, the Department of Energy had these scientists go to the various sites, inform the local residents of the decision, and answer questions that the people had.

At the first location, which was in a western state, the scientists held a meeting in a town hall and adopted the old strategy, *Tell them what you're going to tell them; tell them; tell them what you told them.* The strategy failed miserably. As soon as the scientists announced the decision that this site was a finalist for the nuclear waste repository, the crowd of ranchers and farmers unleashed a firestorm of questions: Why were we chosen? What will happen to our livestock? What will happen to our crops? How safe will it be to drink the water? The scientists tried as best they could to reassure the audience that their decision in no way would affect the ranching and farming that went on in the area. In fact, this place was chosen for that very reason: The geography of the area was such that the ranching and farming would be able to continue without effect. However, the attempt to pacify the crowd came too late in the presentation. Everyone in the town hall was speaking at once, and many in the crowd had stopped listening to the scientists. The ruckus continued with many in the crowd leaving in disgust and those who remained continuing to hold their position of "not in my back yard." When the meeting finally concluded and the scientists walked out to their rental car, they saw that someone had dropped a load of manure on top of it.

Clearly, these scientists had not accounted for the bias of their audience.

Understanding the bias of the audience helps you decide both the strategy and the energy required for a successful argument. For instance, solidifying support with an audience that already leans toward your position or is neutral toward your position does not require nearly the energy that garnering support does from an audience that is antagonistic to the position. For instance, engineers at Morton Thiokol were able to persuade their management that the launch of the space shuttle *Challenger* should be delayed until there were warmer temperatures. However, these same arguments made to NASA later in the day did not succeed. The main reason was that the initial bias of NASA against a delay was much stronger than the initial bias of Morton Thiokol's managers.[9]

Sometimes, the initial bias of an audience is the overriding factor in determining the success of a presentation. Contrast the failed one-on-one presentation of Niels Bohr with Winston Churchill in 1944 with the surprisingly successful one-on-one presentation of Edward Teller with President Reagan in 1982. In Bohr's meeting with Churchill, his purpose was to have Churchill realize the potential nuclear weapons race that Bohr anticipated would follow the Second World War. However, Churchill, already defensive about his decision to relinquish intellectual rights to nuclear weapons, ended the meeting after only twenty minutes and asked Bohr to leave.[10] The purpose of Teller's meeting with Reagan was to persuade him to change the United States nuclear weapons policy of mutually assured destruction to a policy of a strategic defense initiative. Given the resistance in the military to such a change and doubts by other scientists such as Hans Bethe as to the potential of the initiative, such a goal seemed out of reach. However, the receptiveness of Reagan and some of his advisors to an alternative to mutually assured destruction proved to be an ally for Teller. The result of that meeting and a later meeting between Teller and one of Reagan's advisors led to the dramatic shift in nuclear weapons policy in March 1983.[11]

With an antagonistic audience, two strategies should

be considered. One strategy is to define the question up front, but not to give away your results. If those in the audience who are opposed to your results do not know your position, they are much more likely to listen to your arguments. Granted, if their initial bias is strong, you probably will not change their minds by the presentation's end, but you are in a much better position to reduce their vehemence against your position. You might also win their respect.

A second strategy, named the Rogerian strategy for the psychologist Carl Rogers,[12] is to show that you truly understand the opposition's main arguments. In other words, you extend an olive branch to the opposite side by recognizing the strengths of their argument before you begin with a defense of your own. What this olive branch does is to reduce the initial antagonism that the audience has to you and makes them more inclined to listen to your arguments. Such a strategy works well when the goal is not to win the other side over, but to reach a compromise with the other side.

In cases in which you desire to win over an audience antagonistic to your position, do not set your expectations too high. As the physicist Max Planck asserted, "An important scientific innovation rarely makes its way by winning over and converting its opponents—it rarely happens that Saul becomes Paul."[13] Although you might have little success winning over your opponents, using one of the two strategies presented can help you reduce the opposition to your position and perhaps win over those who are neutral on the subject.

Critical Error 4
Losing the Audience at Sea

At the end of the presentation, when the speaker asked for questions, Professor Sigmar Wittig rose and said flatly, 'Sir, I have been listening to your talk for the past fifteen minutes, and I don't believe a word that you have said. In two minutes, using the First Law of Thermodynamics, I can prove that everything you have presented is nonsense.' The speaker turned pale. But I turned paler, because the next day I was to give my presentation, the first of my career, and Professor Wittig was sure to be in attendance.[1]

—Karen Thole

In 1860, James Clerk Maxwell, who is considered the father of electrodynamics and one of the greatest physicists of the last two centuries, applied for a chaired professorship at the University of Edinburgh. He did not get the job. Instead, it went to Peter Guthrie Tait. According to an article in the Edinburgh *Courant*, the reason for Maxwell not getting the position was his lack of skill at speaking.[2] The reasoning of those who made the selection was that whoever taught had to be able to communicate to an audience (the students) that would not know the subject. What made people consider Maxwell a weak speaker? According to one of his students, C.W.F. Everitt,[3] Maxwell prepared lectures that were well organized. He wrote them out in a form that Everitt claimed was "fit for printing."[4] However, soon after beginning to lecture, Maxwell would digress onto a long tangent, filling the blackboard with equations and illustrations, thinking out loud, and surpassing the comprehension of his audience. Maxwell's tangential discussions went on so long that the lecture time would run out, and his original organization would not be presented.

As mentioned in Chapter 1, a major disadvantage of presentations is that the audience does not have the luxury, as they have in a document, to go back and reread a passage. For that reason, an audience can easily become lost. Even when the speaker is careful, the audience can become distracted and fall behind. How many times in a presentation have you started contemplating a connection between the speaker's work and your own work and then snapped back to the presentation, only to discover that the speaker has moved to another topic and that you are unsure what has transpired?

Given the inherent potential for the audience to become lost even when the structure is sound, consider how easy it is for the audience to become lost when the structure is weak. Several instances can arise in a presentation to cause the audience to become lost. One occurs when the presenter gives a presentation that contains gaps in logic, or, figuratively speaking, when the presenter launches a vessel that is not seaworthy. A second instance occurs when the presenter does not clue in listeners about a major change of course in a presentation. A third instance occurs when the presenter drowns the audience in detail.

Launching a Ship That Is Not Seaworthy

When describing the presentations of Niels Bohr, Einstein said, "[Bohr] utters his opinions like one perpetually groping and never like one who believes himself to be in possession of definite truth."[5] C.F. von Weizsäcker claimed that Bohr's presentations reflected the great physicist's way of thinking, which Bohr himself had compared to a Riemann surface. According to von Weizsäcker, the complexity of Bohr's thinking was reflected in his "stumbling way of talking" that "would become less and less intelligible the more important the subject became."[6]

Rather than presenting those subjects with which he was grappling, Einstein chose to present those topics that he felt he understood. For that reason, Einstein came across to audiences as much more lucid and confident than Bohr. This difference between the presentations of Einstein and Bohr raises the question about what engineers and scientists should present. Should they present only what they know to be stone-cold facts? Or should they expand the boundaries and present what they suspect to be the case?

In the latter case, if the ship is not yet seaworthy, as in the claim for cold fusion made by two researchers at a press conference on March 23, 1989, then the presenters could be embarrassed.[7] However, if the ideas prove to be correct, then the presenters stand to receive credit for the bold step. Such was the case for James Watson and Francis Crick when they proposed the double helical structure of DNA. Interestingly, Rosalind Franklin's notebooks from the winter of 1952–1953 reveal that she was very close to finding the structure for DNA.[8] Unlike Watson and Crick, though, she was much more cautious about making jumps.

A scientist who took bold leaps in presentations was Linus Pauling. Pauling's courage (some might say audacity) went well beyond presenting theories that were not fully validated in the laboratory. On several occasions, Pauling presented theories that were, at best, sketchy. In some cases, Pauling was simply wrong, as was the case in his theory that antibodies fastened themselves to antigens by curling up around them.[9] However, many times Pauling was correct or at least close enough that he received credit for the idea. One example was his argument for the chain theory to explain the structure of proteins. That theory went against the cyclol theory, which at that time had a much stronger mathematical basis and was much widely more accepted by the scientific community.

Given the dramatically different results in the examples above, this issue about whether to present something that is not fully validated remains difficult to answer. On the one hand, the safe advice is that you should present only what you know for certain. In doing so, you certainly reduce the risk of embarrassing yourself. On the other hand, one of the advantages of making a presentation is that you can receive feedback from the audience about your work. If you are stuck on a problem, presenting a "straw-man" solution to an audience could trigger a suggestion from the audience that would help you solve the problem. In some situations, you could view presentations as tests for ideas. In my own experience of teaching scientific writing at the national laboratories, sometimes I have tested a piece of advice on an audience at the laboratories. If the advice did not ring true with what the engineers and scientists experienced in their work, I quickly found out.

With this question of whether to limit yourself to cold facts or to include conjecture, much depends upon the audience, the purpose, and the occasion. If you are reporting to an audience in which you cannot afford to stumble, then relying on stone-cold facts makes sense. For instance, if you are a researcher presenting your work to the principal funding organizations in your field, it would not be wise to take large risks. With a more forgiving audience, though, such as the colleagues with whom you have established credibility, taking a chance would probably have fewer consequences. Another variable is how much risk you are willing to take. Linus Pauling risked much, yet reaped much from his risks.

Some situations, such as progress reviews, demand that you present your results even when you do not yet understand them. In such cases, what should be your strategy? A tendency for presenters is to downplay results that they do not understand. However, such a strategy runs the risk of having those results brought to the surface during question periods. After all, during question periods, the audience focuses on those aspects in the presentation that were not understood. A different strategy follows the adage at Dow Chemical: *If you can't fix it, feature it.*[10] In other words, if you do not understand a result, then let the audience know up front that you do not understand it. As with answering the question about what assertions to include in a presentation, answering this question about what reservations to admit in a presentation depends upon the audience, the purpose, and the occasion.

Failing to Warn About Changes in Course

Becoming lost in a sailboat at sea is an uncomfortable experience. The more that you sail, the more tired you become and the less sure you are that you will reach land. Although not nearly as unsettling as being lost at sea, being an audience member lost in a presentation is also an uncomfortable experience. You are not sure where you are in the presentation. Are you at the end of the beginning, the beginning of the middle, or the middle of the middle? Because the speaker has not kept you on track, you concentrate more and listen for clues from the speaker. The harder that you listen for clues on where you are, though, the more tired you become. Because of an audience's frustration that arises when they do not know where they are, speakers should be sensitive to keeping the audience on course, especially when the presentation changes direction.

As was shown in Figure 3-6, certain places in a presentation naturally have a shift in direction: from the beginning to the middle, from the first division of the middle to the second division, and so on. These transitions are important in helping the audience to remain on course. The transitions occur on two different levels. On the first level are the transitions between the beginning and the middle and between the middle and the ending. The transition between the beginning and the middle is important for allowing the audience to assign details to each of the major divisions of the presentation.

As mentioned earlier in the chapter, the shift between the middle and the ending is important for emphasis reasons. Because an audience will sit up and pay more attention during the ending of a presentation, the speaker should make it clear when the ending is upon them. Unfortunately, many inexperienced speakers do not clue in the listeners that the ending is upon them. Rather, these speakers race into port and abruptly ask the audience, "Any questions?" The audience, unprepared for the ending, has lost the opportunity to think about the work from a big-picture perspective and to assess which aspects of the presentation were most important. This mistake diminishes the chances that the audience will catalogue and remember the important details of the presentation.

A second level of transitions occurs between each segment of the middle. Ideally, middles are broken into two, three, or four divisions. For the audience to pace themselves in the middle, the audience has to know where the speaker is: in the first division, in the second division, and so on. For that reason, the speaker should make clear those transitions between each division of the middle.

So how do you make the transitions for those shifts? Several ways exist. The most straightforward occurs in speech. For instance, in moving from the first section of a middle to a second section, the speaker could state that transition explicitly: "That concludes what I wanted to say about the building stages of Hawaiian volcanoes. Now I will consider the declining stages." For the transition between the middle and the ending, the speaker can use phrases such as "in summary" or "to conclude this presentation."

An additional way to make transitions between segments of a presentation is a change in slides. Figure 3-11 shows the mapping slide and three transition slides from a presentation on using composite materials in the bipolar plates of fuel cells. The mapping slide for the presentation includes a key image for each of the three divisions of the presentation's middle. Each image is then repeated as an icon on all the slides for that image's corresponding division. Shown in Figure 3-11 is the first slide of each division. Interestingly, the first slide for the third division is itself a mapping slide that outlines three topics for that division: polarization curves, power curves, and visual inspection. By showing these icons throughout the presentation's middle, the speaker continually reminds the audience where they are in the middle—in the first division, in the second division, or in the third division.

Delivery also provides excellent ways to signal a transition between sections of a presentation. One signal is a pause. In a presentation, a pause is not initially taken as a sign that the speaker is lost. Rather, the audience assumes that the speaker is collecting his or her thoughts. Moreover, a pause in a presentation, like the white space after a section in a document, allows the audience to collect their thoughts as well. After all, in a successful presentation, everyone is working: the speaker to deliver the details and the audience to sort, synthesize, and analyze those details. Well-placed pauses allow that sorting, synthesis, and analysis to occur.

Other aspects of delivery can signal a transition in a

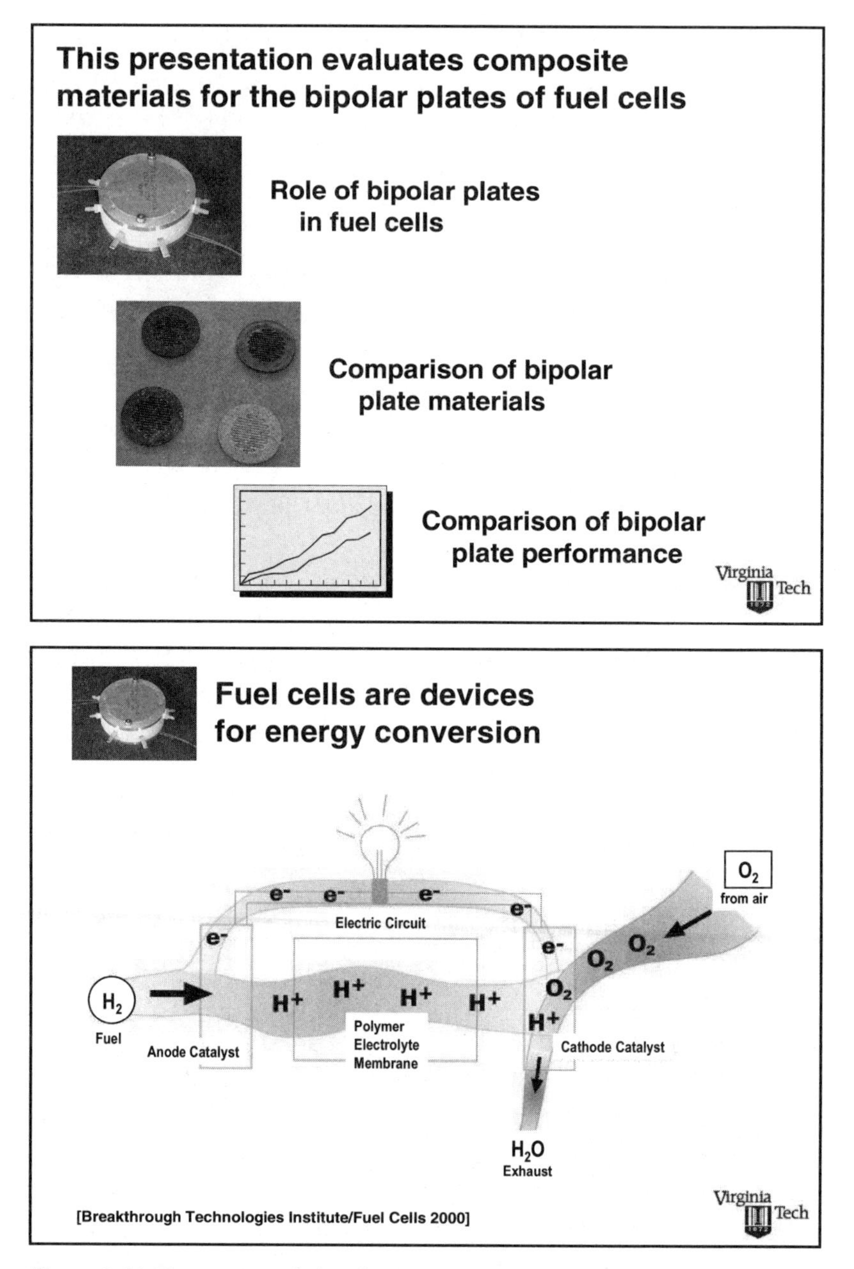

Figure 3-11. Transition slides from a presentation on using composite materials in the bipolar plates of fuel cells.[11,12] Top left is the mapping slide with key images. Bottom left is the initial slide from first division of middle. Top right is the initial slide from second division of middle, and bottom right is the initial slide from third division of middle.

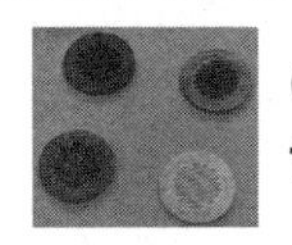

Composite materials are ideal for bipolar plates

Advantages
- Easy to shape
- Light in weight
- Resistant to corrosion

Disadvantages
- Low conductivity
- High cost (at present)

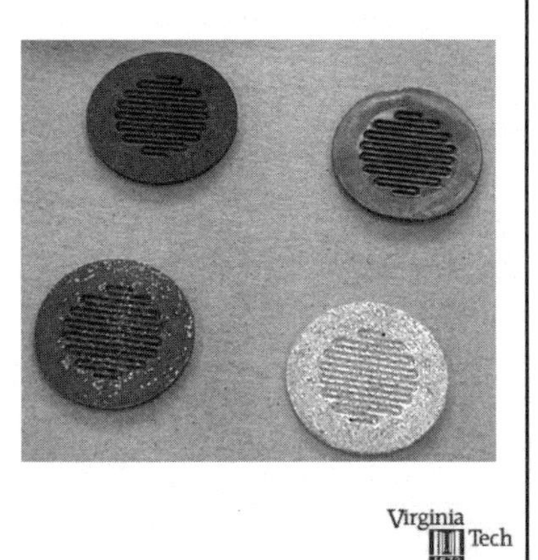

Virginia Tech

There are three methods for bipolar plate evaluation

Polarization Curves

Power Curves

Visual Inspection

Virginia Tech

Figure 3-11 (Continued).

presentation. One signal is the gesture of holding up one, two, or three fingers when mentioning that you are now covering the first, second, or third division of the middle. Another signal is the raising (or lowering) of your voice when beginning a new section. Yet another is moving closer to the audience at each main division of the presentation. A fourth is adopting a certain position, perhaps behind the podium, at a new division. In his series of "Messenger Lectures" given at Cornell University,[13] Richard Feynman returned to the podium, paused, and glanced at his notes between each segment of his presentation. These repeated motions of delivery provided a clear signal to the audience of the lecture's divisions.

Drowning the Audience in Detail

Perhaps the most common way that speakers lose audiences in presentations is that they drown their audiences with details. When you effectively present your work, you do not present everything about your work. Rather, you select those details that allow the audience to understand the work, and you leave out details that the audience does not desire or need. Effectively presenting your work also means that you sort details so that the audience is not faced with a long laundry list that has to be catalogued and synthesized on the spot. Finally, effectively presenting your work means that you provide a hierarchy of details so that the audience knows which details to hang onto and which details to let go in case they are overwhelmed.

One of Niels Bohr's weaknesses as a presenter was that he did not leave out details. Bohr's passion for completeness overshadowed his audience's need for clarity. In his Nobel Prize speech, Bohr's passion for completeness can be seen in the long lengths and complex structures of his sentences:

> The present state of atomic theory is characterized by the fact that we not only believe the existence of atoms to be proved beyond a doubt, but also we even believe that we have an intimate knowledge of the constituents of the individual atoms. I cannot on this occasion give a survey of the scientific developments that have led to this result—I will only recall the discovery of the electron toward the close of the last century, which furnished the direct verification and led to a conclusive formulation of the conception of the atomic nature of electricity which had evolved since the discovery by Faraday of the fundamental laws of electrolysis and Berzelius's electrochemical theory, and its greatest triumph in the electrolytic dissociation theory of Arrhenius.[14]

In the second sentence, Bohr's mentioning of one detail (the discovery of the electron) provoked him to bring a second detail (electricity) into that same sentence. Still in that same sentence, mentioning electricity provoked him to bring in yet more details: electrolysis, electrochemical theory, and electrolytic dissociation theory. One problem with the addition of those details was that the audience was not prepared for them, because earlier in the sentence Bohr had promised to recall only "the discovery of the electron."

Contrast that dense opening of Bohr's Nobel speech to the clear opening of the Nobel speech that Christiane Nüsslein-Volhard gave:

> In the life of animals, complex forms alternate with simple ones. An individual develops from a simple one-celled egg that bears no resemblance to the complex structure and pattern displayed in the juvenile or adult form. The process of embryonic development, with its highly ordered increase in complexity accompanied by perfect reproducibility, is controlled by a subset of the animal's genes. Animals have a large number of genes. The exact number is not known for any multicellular organism, nor is it known how many and which are required for the development of complexity, pattern, and shape during embryogenesis. To identify these genes and to understand their functions is a major issue in biological research.[15]

In this opening, Nüsslein-Volhard quickly brings the audience from the general (life of animals) to the specific (identifying genes) without leading the audience on unnecessary side paths.

Another way that speakers drown audiences with details is to present lists that are too long. Audiences remember lists of twos, threes, and fours. Having lists with six, seven, eight, even nine items is overwhelming for most listeners. The effect is that many in the audience will give up, their eyes will glaze over, and they will daydream about their own work. So what happens if you want to present a list of eight items, such as the eight stages of a Hawaiian volcano? In this case,[16] you can break the list up into smaller lists that the audience can catalogue more easily:

- Building Stages
  - Explosive Submarine Stage
  - Lava-Producing Stage
  - Collapse Stage
  - Cinder-Cone Stage
- Declining Stages
  - Marine and Steam-Erosion Stage
  - Submergence and Fringing-Reef Stage
  - Secondary Eruptions and Barrier-Reef Stage
  - Atoll and Resubmergence Stage

Yet a third way that speakers drown audiences with details is that speakers fail to assign a hierarchy to details so that the audiences can decide which details to hang onto and which details to leave behind. Not all listeners in a presentation will comprehend and retain the same number of details. For that reason, speakers have to be careful to make sure that the audience knows which details are more valuable. Richard Feynman in his introductory lectures on physics at Caltech recognized this predicament. On the one hand, he wanted to challenge the best listeners by presenting tangential details that would expand their thinking, but on the other he did not want to lose those who were struggling to keep up with the main points. What Feynman did was to "write a summary of the essentials on the blackboard"[17] at the beginning of each lecture. For his situation, Feynman adopted a strategy of telling the audience up front what the most important points were, and for this situation, that strategy worked well.

Just as Feynman found a way to reveal the hierarchy of details, so should each presenter find a way to let the audience know which details are most important. One way is repetition: mentioning the detail in the beginning, repeating it in the middle, and then repeating a second time at the end. Another way is similar to what Feynman did: placing key results and images onto the slides and having less-important details mentioned only in the speech. Yet a third way to emphasize information is in the delivery: pausing before an important point; raising the voice or, often more effective, lowering the voice; or stepping closer to the audience so that they sense a difference in the emphasis of the presentation.

Chapter 4

Visual Aids: Your Supporting Cast

In his lectures Faraday communicated not only through words but by his deft manipulation of nature. As a virtuoso he could make operations of nature manifest before the audience's eyes. His audience thus both heard about the laws of nature and also saw them in action. It is significant that in an early letter Faraday had placed the power of the eye far above that of the ear in conveying ideas clearly to the mind. Thus observable experiments played a central role in Faraday's didactic repertoire. For the audience the immediacy of this experience proved one of the most potent aspects of Faraday's lectures.[1]

—Geoffrey Cantor

Having a supporting cast of visual aids has many advantages, but almost as many pitfalls. One major advantage is that a strong supporting cast of projected slides, demonstrations, and pass-arounds can convey images, sounds, textures, tastes, and smells much more effectively than spoken words can. With a projected slide, for example, you can precisely and efficiently show the arrangement of a complex image such as the gas turbine engine depicted in Figure 4-1. Moreover, in a demonstration, you can provide the audience with the simulated roar of the *Parasaurolophus* dinosaur.[2] In addition, with pass-arounds, you can allow audiences to feel a snake's skin, to experience the garlic taste of dimethyl sulfoxide, or to

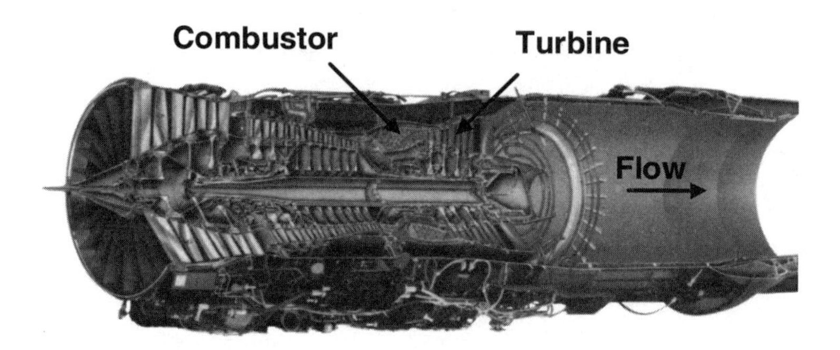

Figure 4–1. Cutaway of a gas turbine engine.[3] Fuel is burned in the combustor and flows downstream to drive the turbine. An interesting problem is that the temperatures of the combustion gases coming out of the combustor are much higher than the melting temperature of the metal vanes in the turbine. What prevents those vanes from melting is film cooling provided by bleed air from the compressor.

sniff the odor of acetone (which on a patient's breath is a symptom of diabetes). By incorporating a supporting cast of visual aids, you allow the audience to experience those sensations. That experience is important, because people remember experiencing the five senses much more readily than they remember descriptions about those sensations. Close your eyes for a few moments and think about your first year in grade school. What comes back to you? Rather than words you read or heard, you are more likely to remember specific sensations: a yellow school bus rounding a thick stand of pines, the slap of wooden desktops, the porous texture of cinderblock walls, the taste of fish sticks doused in ketchup, the scent of new spiral notebooks. If you do recall exact words, you probably also recall sensations from the scene in which those words were spoken.

Another advantage of visual aids is that they can offer variety and beauty to your presentations. People can listen only so long before they need a break. Visual aids can provide that break. In a way, visual aids act as another presenter on stage. Just as experiments gave

Michael Faraday's presentations a distinct look and feel, so too can different visual aids give your presentations a unique look and feel.

A third advantage is that well-crafted visual aids reveal that the speaker has put much effort into the presentation. That effort can make a good impression on an audience. Many audience members sense that if the presenter put so much energy into the presentation, then the presentation might be worth their attention.

What are the drawbacks of employing visual aids? One drawback is that a visual aid that fails can undercut the speaker's authority and undermine the entire presentation. In one sales presentation, for instance, an engineer was demonstrating a pump that used graphite rather than oil as the lubricant. Unfortunately, during the demonstration, the pump began emitting a thick cloud of black smoke. Although the engineer quickly turned off the pump, the audience could see, hear, and smell that the pump had burned itself out. Needless to say, no pumps were sold that day.

Another drawback is that creating visual aids of high quality is an intimidating venture for many scientists and engineers. Many scientists and engineers feel like novices wielding a camera or working on a graphics program. Fortunately, digital cameras have made it much easier not only to capture images and films, but also to preview those images and films as soon as they are taken. Also fortunate is that the computer programs for creating graphics and films have become easier to use. Moreover, for those difficult photographs, drawings, diagrams, films, and models, many institutions have professionals to provide assistance.

Yet another drawback of incorporating visual aids is that poorly designed visual aids can raise more questions than answers for an audience. More than once I have walked away from a presentation thinking that the presenter would have done better to turn off the projector and just speak to the audience. In those presentations, what was projected simply was not worth the break in the presentation's continuity. In one instance, an engineer who was chosen to lead off a research group's lecture series began the inaugural presentation by projecting a six-by-six table that showed the different divisions of his research. One problem with this table was that the lettering on the table was much too small for anyone to read (and I was in the front row). Worse yet, the six-by-six array intimidated everyone in the room. Was this person going to talk about all thirty-six divisions? Unfortunately, he did, with equally complex visuals for each. By the way, that first presentation of the lecture series was the last presentation of the lecture series. No one wanted the series to continue.

When designing visual aids, you should consider their effect on the audience. For instance, when you project a slide on the screen, the audience breaks its eye contact with you and looks at that screen. The audience tries to discern what is projected and how this projection fits into the scheme of the presentation. When slides work well, the slides orient the audience quickly. The slides also provide a perspective on the information that could not be achieved with simple speech. Unfortunately, too many slides rely on phrase headlines, which often do not orient well, and bulleted lists, which quickly place audiences in trances. One reason that so many slides have this format is because that format is the default of programs such as Microsoft's PowerPoint. This book does not mean to suggest that you should not integrate computer programs such as PowerPoint into presentations. What this book advocates, though, is that scientists and engineers critically evaluate the defaults suggested by such programs. For instance, my wife uses PowerPoint to create effective slides in her engineering presentations.

However, she does not use the program's defaults for the typography or layout of her slides. Rather, she uses the program to create her own slide designs that communicate her work to her audiences.

Presenters have different types of visual aids from which to choose: projected slides, posters, writing boards, films, demonstrations, models, handouts, and passed objects. Table 4-1 summarizes the advantages and disadvantages of each. Except for posters, these types are discussed in more detail in this chapter. Posters, which are used for a special type of presentation, are discussed separately in Appendix B.

Table 4-1. Advantages and disadvantages of different types of visual aids.

Type	Advantages (+) and Disadvantages (–)
Projected Slides	+ can effectively show images + can effectively emphasize key details - are boring if no images are included - are overwhelming if too many details are included
Posters	+ allow audience to control pace + allow for one-on-one exchanges with speaker - are difficult to read in crowded hallways - are overwhelming if too many details are included
Writing Boards	+ are good for derivations (the pace is natural) - force presenter to turn away from audience - are slow for detailed drawings - are difficult to read if handwriting is poor
Films	+ are effective for showing dramatic changes + are effective for conveying sounds - cause the audience to focus solely on screen - must meet high expectations of the audience
Demonstrations	+ are effective for engaging the audience + are effective for incorporating sounds and smells - can fail
Models	+ are effective for showing three dimensions - are ineffective unless large enough for audience to see
Handouts	+ ensure that audience leaves with the message - can be distracting if distributed too early
Passed Objects	+ allow audience to touch, smell, and taste - can be distracting if audience is large

Projected Slides

In conference presentations, projected slides such as overhead transparencies and computer projections are commonly used, and with good reason. Such slides, when well designed, can greatly increase the retention of details. Most people remember on average about 10 percent of what they hear and 20 percent of what they read. However, with well-designed slides, the retention of an audience that is both hearing and seeing the information can increase to 50 percent.[4]

Unfortunately, many presenters do not use the types of designs that would increase that retention to 50 percent. As is discussed in Critical Error 5, many presenters place so many details on their slides that audiences do not even attempt to read those slides. Also, as discussed in Critical Error 6, other presenters simply create designs of bulleted list after bulleted list that place audiences into trances. In both Critical Error 5 and Critical Error 6, this book focuses on slide designs that are best for communicating scientific information to the audience. Interestingly, the designs proposed in this book challenge many of Microsoft PowerPoint's defaults, which perhaps Microsoft chose for communicating more general information.

While this book later examines the difficult question of how best to design projected slides, this book now considers only the more straightforward question of which type of slide (overhead transparency, computer projection, or 35-mm slide) is best to use.

Overhead Transparencies. Using overhead transparencies has two main advantages over using computer projections or 35-mm slides. First, an overhead projector is almost always available. That availability is a distinct advantage if you are travelling to make presentations and are not confident about the equipment on hand. Second, using overheads gives you the flexibility to discard transparencies or to switch the order of the transparencies in the middle of a presentation. The ability to remove projections is valuable if the presentation has a fixed time limit. To stay within the time limit, you simply do not show those overheads that present secondary information. The ability to reorder overheads is valuable if the presentation's format allows the audience to interject questions, because you can quickly bring forward those overheads that address the questions.

One disadvantage of using overheads is that editing the overheads is relatively difficult (compared with editing a computer projection), especially if you are travelling. A second disadvantage, which actually should not be a disadvantage, is that the equipment seems to pose a challenge for many speakers. Granted, five minutes of practice is all that is needed to learn how to turn on the projector, to position a transparency such that it projects properly, to focus the image, and to see where to stand so that the image is not blocked. However, given the misuse of overhead projectors by so many presenters at conferences, lectures, and meetings, this second disadvantage cannot be discounted.

Computer Projections. Computer projections are an excellent way to display beautiful graphics and important words. In addition, these projection systems are smaller in size and lighter to carry than overhead projectors. As the costs of these systems come down, these systems will replace overhead projectors as the standard means of projecting slides at conferences. The advantages of these systems are numerous.

First, with a computer projection, the presenter can

modify the presentation right up to the moment of the presentation. Second, the presenter can build slides (as shown in Figure 4-2) to allow the presenter to work in a number of details that would intimidate the audience if shown all at once. Moreover, this building replaces the awkward practice of the presenter sliding a sheet of paper down an overhead as the presenter goes through the different points. This practice is awkward because so often the paper falls off if the speaker moves away from the projector (by the way, a better practice with overhead transparencies is to use self-stick sheets that do not slide off). A third advantage of a computer projection is that the presenter can easily incorporate a film into the presentation. These films can arise from computer simulations or from movies captured on digital cameras.

One disadvantage of a computer-projected presentation is that the speaker is locked into the order established at the beginning. In other words, if the presentation has nine slides in the middle, then the speaker has to show all nine before getting to the conclusion slides. In a presentation that has a fixed time limit, this locked order prevents the speaker from discreetly leaving out a slide. Granted, the speaker can scroll through the slides that he or she does not have time to cover in more depth, but then the audience realizes that they have missed something.

Another disadvantage of computer-projected presentations is the uncertainty of whether the computer will work as expected with the projector. While many institutions now have projectors, there is no guarantee that your computer will work with their projector, especially if the institution's projector is one of the early projectors, because many of these earlier types require special software to be installed on your computer. You can increase the odds by using your own projector and computer, but all that equipment can be cumbersome to transport.

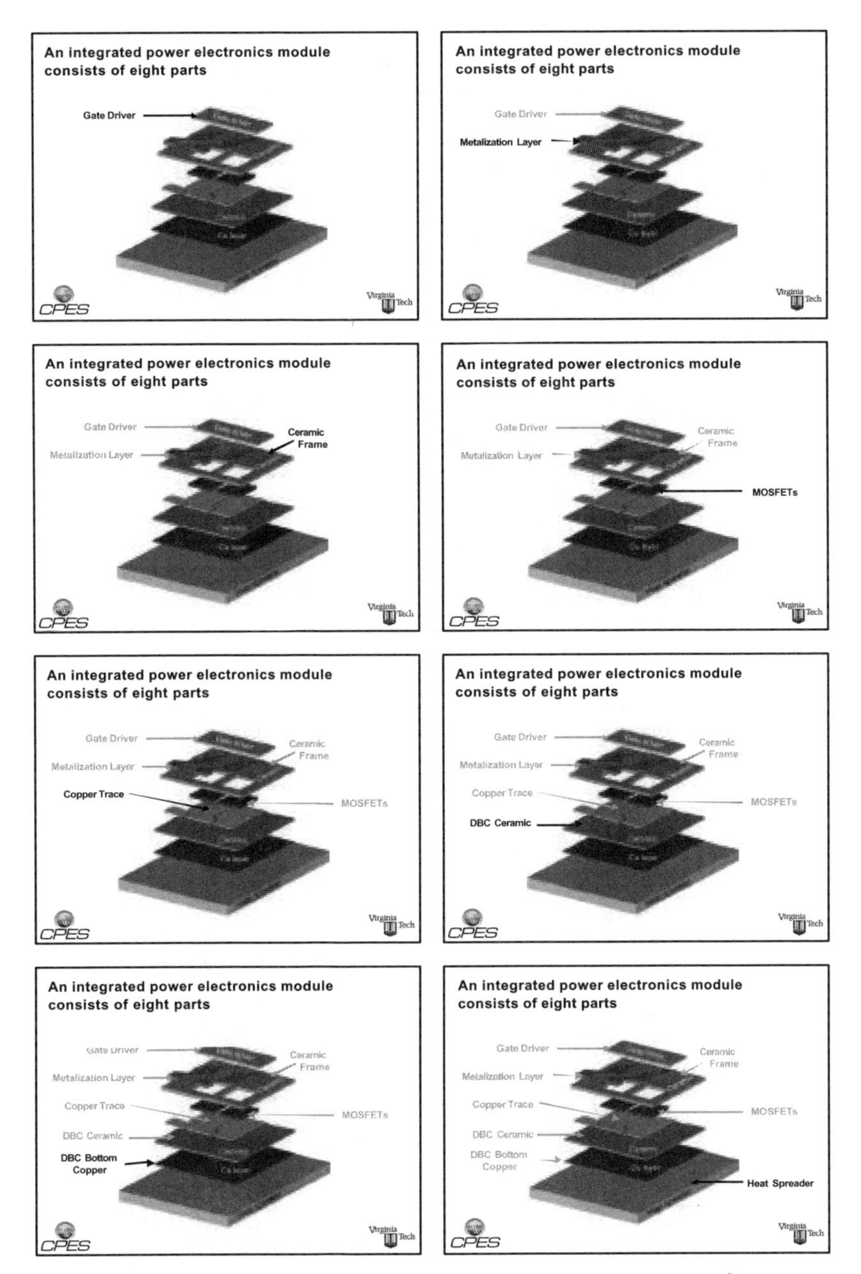

Figure 4–2. Sequence of a building slide that shows parts of an integrated power electronics module.[5] The building allows the audience to comprehend one segment of the design before the next is shown.

35-mm Slides. Once a fixture in scientific presentations, 35-mm slides are now less commonly used than overheads or computer projections. The main advantage of 35-mm slides is the high resolution of the projected images. Computer projectors are matching that quality, though. Another advantage of 35-mm slides is that the projection equipment is relatively easy to transport.

The drawbacks of 35-mm slide presentations, though, are numerous. The first is the same drawback of computer projectors: the inability to reorder the slides to respond to a question or to skip readily over slides to make a time deadline. Yet another drawback for a 35-mm slide presentation is that the lights must be so low that the speaker cannot make good eye contact with the audience. A final drawback, and one that will make them obsolete in scientific presentations, is that it takes significantly more time to create a 35-mm slide than it does to create an overhead or a computer projection.

Writing Boards

In the mid-nineteenth century, writing boards were a novelty. Institutions would boast about having the only chalkboard in a 100-mile radius. Just as people a few years ago were excited by the novelty of seeing a presentation that used a computer projector, people in the nineteenth century were excited about seeing a presentation that used a chalkboard. Also, just as recent audiences have developed higher expectations for slides from computer projection systems, so too did audiences who witnessed the advent of the chalk board increase their expectations about the its uses. In other words, as audiences saw presenters use writing boards in more sophisticated ways (rulers to make lines, chalk tied to string to make circles, colored chalk to show textures), the expectations of those audiences increased.

In essence, writing boards are a form of slides in which the presenter creates images and produces words or equations as the presentation occurs. For that reason, the pace of the presentation is dictated by the pace at which the presenter places information on the boards. This constrained pace can be an advantage in a mathematical derivation. In fact, one of the best uses of using a writing board is for deriving equations.

Writing boards, though, have their disadvantages. One disadvantage of writing boards is that the speaker has to turn away from the audience to write on the board. A second disadvantage of writing boards is that the quality of the images and lettering depends upon the ability of the presenter.

Despite the disadvantages, many scientists and engineers have become adept at using writing boards. For instance, Richard Feynman was considered a master of board work. He planned many of his lectures such that the writing for the lecture began in one corner and ended in the opposite corner.[6] Another person who was adept at using writing boards was Ludwig Boltzmann. According to Lise Meitner,

> [Boltzmann] used to write the main equations on a very large blackboard. By the side he had two smaller blackboards, where he wrote the intermediate steps. Everything was written in a clear and well-organized form. I had frequently the impression that one might reconstruct the entire lecture from what was on the blackboards.[7]

Just as users have become more adept at working with writing boards, so too have writing boards become more sophisticated. For instance, some boards, called smart boards, are linked to computers to allow information to be stored and for copies to be made.

Films

The principal advantage of a film is to show dramatic changes over a relatively short period of film time. Even lengthy periods of real time can be captured, as long as the film time is short enough and the changes dramatic enough to maintain the audience's attention. A good example of a technical film is Robert Ballard's computer simulation that demonstrated the sinking of the *Titanic*.[8] In this simulation, the three hours that it took the Titanic to sink was condensed to about twenty seconds, a time long enough to show the stages of the sinking, but short enough to maintain audience interest.

Another strong example of a film is Gregory Pettit's morphing of an airplane wing to produce different flight conditions. Shown in Figure 4-3 are two frames from this film. The wing achieves its morphing, or reshaping, through a special material and construction. Shown in the bottom left portion of each frame is the physical reshaping of the wing. The flight conditions, such as altitude and airspeed, that the morphing produces are shown graphically in the top portions of the frames. Shown in the bottom right portion are those same conditions presented on an instrument panel.

The best technical film that I have ever witnessed was a NASA film shown by the astronaut Ellen Ochoa.[9] This film, which presented the *Atlas-3* mission of the space shuttle, spanned the entire middle of Ochoa's presentation. As you would expect, the film followed a chronological strategy for the mission, beginning with the preparation for the shuttle launch and concluding with the shuttle's landing. One aspect that made this film stand out was NASA's incredible photography; the launch itself was shot from many angles. Another aspect was the careful editing by NASA's staff to capture the most important details of the mission and to leave out unnecessary scenes.

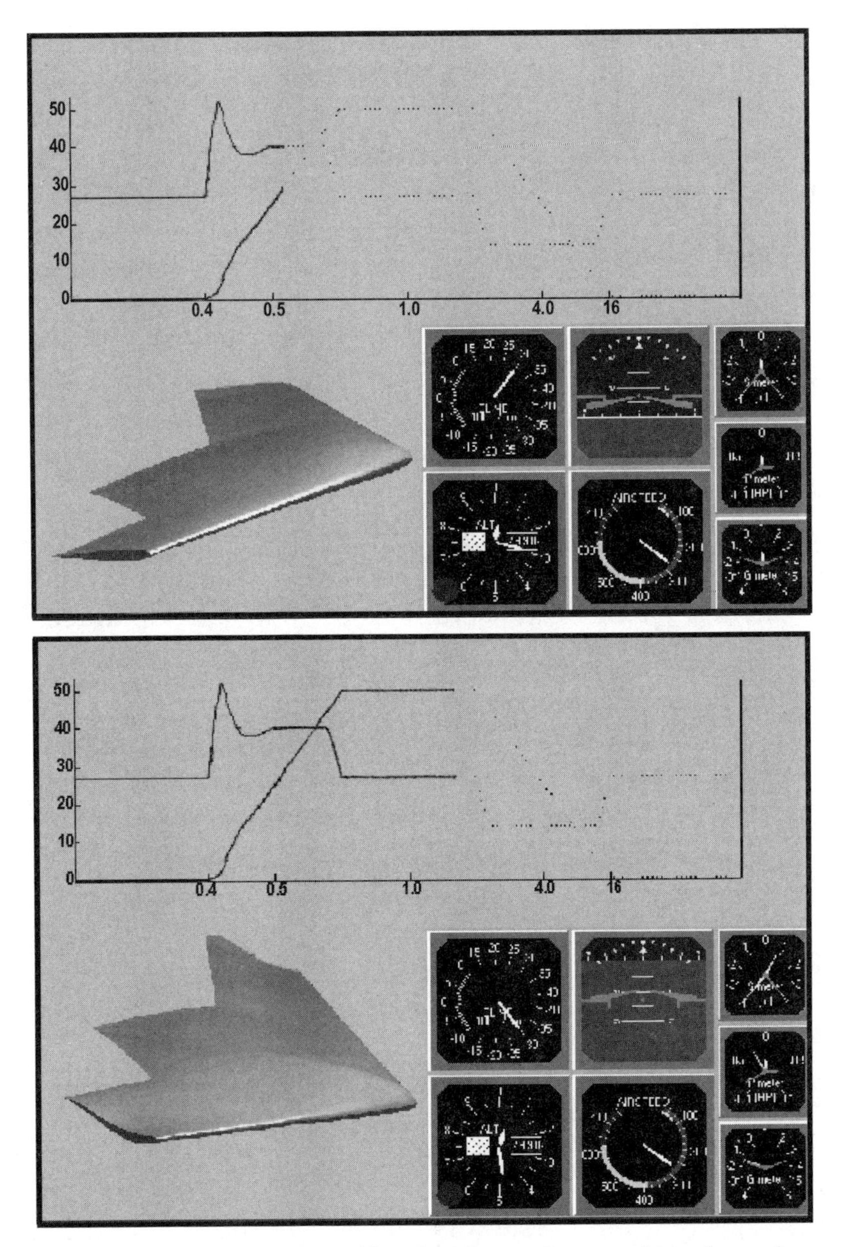

Figure 4–3. Two frames from film that shows the morphing (or reshaping) of an airplane wing as it flies through a desired course.[10] The upper part of each frame shows a plot of both altitude and airspeed. These variables as well as others are shown on instrument panels in the bottom right portions of the frames. The bottom left portions show the reshaping of the wing.

A third aspect that made the film stand out was how well Ochoa narrated the film in a live performance. Clearly, she had practiced this narration, because her timing kept pace with all the activities of the film: the preparations, the launch, the duties on the shuttle, the experiments that the astronauts performed, and their work on the international space station. Although it was clear that Ochoa had practiced her speech, the speech never seemed canned. Her observations, her insights, and her occasional humor sounded spontaneous. Ochoa's presentation was one of the those wonderful, but infrequent, occasions in which the audience wanted the presentation to go longer.

The worst technical film that I ever saw was in an engineering design presentation at a major university. On hand in the audience were the project sponsors from a semiconductor company, professors who had supervised the presenting students, and graduate teaching assistants. The presentation, given by three senior undergraduates dressed in handsome suits, incorporated an overhead projector in the middle of the room and a large television monitor on top of a cart in the room's corner. Except for the film, the presentation was a success—the organization was logical, the speech informative, and the overhead slides supportive. The film, though, was an inappropriate mixture of content and medium.

This presentation occurred in the late 1980s when video cameras were first appearing on the market, and the presenters were proud of the camera that one of them had apparently just purchased. During the first part of the presentation, the presenters mentioned more than once that a film would be presented. Given the number of times that the presenters mentioned the film and the excitement with which they spoke of the film, the audience's expectations were high. When the time for the film arrived, one of the presenters dimmed the lights, and another presenter turned on the video recorder. The video showed a table with a silicon wafer attached to two alligator clips, a digital multimeter, and a voltage source. The audience waited for something to happen, but nothing did. After a few long seconds, the presenter at the podium instructed the presenter by the video recorder to hit the fast-forward button. The presenter did, and everyone stared at a blank screen for a moment. When the presenter released the fast-forward button, the same scene was still there. After several long seconds, the presenter hit the fast-forward button again, and the screen went blank. Again, when the presenter released the fast-forward button, the same scene was there. More long seconds ensued, the industrial sponsors nervously tapped their pencils, and the professors sighed deeply.

Finally, there was some movement on the screen as one of the students, dressed in slacks and a sports shirt, went to the table with the experiment and apparently turned on a voltage supply. Although lights on the digital multimeter flickered, the readout of the digital multimeter was much too far away for the camera to pick up. Recognizing this point, the presenter by the podium (apparently the one who had purchased the camera) stated the reading of the multimeter.

Sensing that the audience was bored by the film, the presenters continued their presentation with the overheads. Unfortunately, they left the video recorder running. For a long time, it showed only the table with the experiment, but then the scene changed—now to a party at someone's house. All the presenters were there, but instead of wearing professional attire, they were dressed in jeans and T-shirts. One wore a sombrero. Although there was no sound on the television, apparently music was playing at the party because all three presenters were dancing.

The presenters did not realize that a party was playing on the television screen behind them. They were busy showing slides and explaining the changes that they had made to their silicon disk. The occurrence of the party, though, was not lost on the industrial sponsors or on the professors, who were red-faced and trying to signal the presenters about the television screen. After twenty long seconds of dancing and beer swilling on the television, one of the professors stated in a barely restrained voice to turn off the video recorder.

This film failed for two reasons. First, the film added nothing to the project because there were no dramatic changes for the film to capture. Another reason that this film failed was that the presenters had not rehearsed the integration of the film into the presentation. Had they practiced even once including the film, they would have realized the importance of when to begin the film and when to end it.

Demonstrations

The advantages of incorporating demonstrations are numerous. Demonstrations can reveal dramatic changes in an exciting fashion. Demonstrations can also reveal more than just images. For instance, in the simulation of lightning strikes to grounded and ungrounded buildings, scientists at the Deutsches Museum in Munich produced both beautiful paths of light, such as is shown in Figure 4-4, and frighteningly loud claps.

For demonstrations to be successful, though, several criteria have to be met. First, the audience has to be able to experience the event, either through sight, sound, or some other sense. Second, the timing of the demonstration has to be such that there is little dead time; otherwise, the audience will become impatient. Third, the demonstration should succeed. Although a deft presenter can recover from a demonstration that goes awry, most of the time when the demonstration fails, the presentation suffers.

Figure 4–4. Lightning demonstration at the Deutsches Museum in Munich.[11] In this demonstration, a lightning bolt strikes a church that is not well grounded. Because the church is not well grounded, a second stroke occurs between the church and a nearby house.

Although Critical Error 7 includes several examples of experiments that do not succeed, the successes experienced by Michael Faraday, Ivan Petrovich Pavlov, and Nikola Tesla provide testaments to the power of including experiments in presentations. All three worked hard to make sure that their presentations succeeded. They practiced and practiced. For a presenter, the decision on whether to use experiments comes down to a balance between the payoffs of audience understanding and the risks of the experiment failing.

Richard Feynman, who served on the presidential committee to investigate the explosion of the space shuttle *Challenger*, performed an interesting demonstration in 1986. Frustrated by the lack of attention being given by the committee to the O-ring design on the solid rocket boosters, Feynman caused all eyes to focus upon him when seated at his chair in a committee meeting he demonstrated how O-ring rubber loses its resiliency when subjected to low temperatures. Such low temperatures were present on the morning of the fateful launch. In his experiment, Feynman simply dipped a clamped O-ring into a glass of ice water, removed the O-ring, and then tried to show that the rubber of the O-ring did not spring back to its original shape when the clamp was removed. Although the audience could not really feel that the O-ring did not snap back, their own familiarity with how rubber stiffens in cold temperatures and the lack of anything else of interest going on in the committee meeting caused that demonstration to make national news.[12]

Models, Handouts, and Passed Objects

Three additional supporting aids are models, handouts, and passed objects. Display models, such as the molecular models held by Linus Pauling as shown in Figure 4-5, are another type of prop that is used in scientific presentations. The main advantage of such models is their effectiveness at showing three-dimensional structures. Another advantage of such models is how well they reveal colors and textures. For that reason, models are often used in presentations that propose architectural designs for buildings. One disadvantage of such models is that they cannot be viewed well by large audiences. Another disadvantage of such models is that they take much time to produce. Yet a third disadvantage is that they are difficult to transport.

In many presentations, the speaker wants to make sure that, at the end, the audience walks out the door with key information. For such presentations, presenters often hand out copies of projected slides. An interesting

Figure 4–5. Linus Pauling holding three-dimensional models of molecules during a presentation (courtesy of the Archives, California Institute of Technology).[13]

question arises with handing out copies of projected slides: When do you distribute the copies? If you distribute the copies before the presentation, the audience has the opportunity to make notes while you speak. A disadvantage, though, is that the audience might ignore your pace for the presentation and look ahead. In so doing, some in the audience might ask questions about information that you have not yet covered in the presentation. That can be distracting. If you hand out the presentation slides at the end, a disadvantage is that the audience loses that opportunity to place notes beside the respective slides. An advantage, though, is that you have the opportunity to surprise the audience with images or results.

In deciding when to pass out handouts, you should carefully consider the content of your presentation. If you do not want the audience reading ahead, then you should wait until the presentation's end to pass out the copies. However, if you decide to pass the copies out at the end, you should let everyone know up front that they will be receiving a handout. Otherwise, they might spend their energies frantically making unnecessary notes rather than paying attention to you.

Yet another type of a demonstration is a passed object. With small audiences, objects that are passed around are often effective. In addition to revealing three-dimensional shapes, these objects can reveal textures, smells, and tastes. The larger the audience, though, the less effective passed objects are, because many in the audience do not experience the model until long after its introduction. Another problem with a passed object is that the object can be something of a distraction. When an audience member receives the object, that person is torn between examining the object in detail or listening to the presenter. Another distraction is that some people hold onto the objects for an inordinately long time, which unsettles those waiting for the object. Passed objects work best when the audience does not feel rushed to examine the object and can examine the object relatively soon after its introduction.

Critical Error 5
Projecting Slides That No One Reads

People are not listening to us, because they are spending so much time trying to understand these incredibly complex slides.[1]

—Louis Caldera
Secretary of the Army

On a regular basis, managers from the different national laboratories trek to Washington D.C. to present their work and to propose new research. During the 1970s and early 1980s, many managers, including those from Sandia National Laboratories, where I worked, held a cavalier attitude about making presentation slides for these meetings. In fact, many Sandia managers prided themselves on waiting until the plane ride to begin creating their slides, which they wrote by hand. In effect, these managers considered themselves researchers, not artists, and therefore did not have time to worry about such trivialities as the design of presentation slides.

One year, though, managers from a competing laboratory, Lawrence Livermore National Lab, broke from this tradition and quietly put much effort into the creation of their slides. Their reasoning was that the projected slides were important for the success of the presentations and that these presentations were important for the success of their proposals. After much deliberation, Lawrence Livermore came up with a specific format that they felt could best communicate technical information to a wide audience.[2] Beginning in the upper left corner of each slide, Lawrence Livermore inserted a short sentence headline, rather than a phrase headline, that stated the main point of the slide. The assertion made by this headline was then supported below by images (at least one image for each slide) and by short groupings of words. At that year's meeting, Lawrence Livermore's presentations were very well received—so well received, in fact, that managers from Sandia vowed never to be shown up again.

When slides are chosen to communicate the images and results of a scientific presentation, their design becomes important for the success of that presentation. Typically, as soon as a slide is projected, the listener shifts attention from the speaker to the screen. When the slide has words that cannot be read, the listener is distracted with the question of what those words are. Likewise, when the slide does not quickly orient the listener, the listener is disoriented, wondering what the point of this slide is. If the presentation does not allow for questions or if the listener is not confident enough to ask a question, then these questions fester in the listener. Given these two distractions, presenters should strive to design slides that are easy to read and that quickly orient the audience.

Despite the importance of designing slides that are easy to read and that orient the audience quickly, many presenters appear to have designed the slides with the opposite intention. For instance, in one recent invited lecture, the presenter used a thin serif font (Garamond) that was hard for the audience to read, even for those sitting in the front row. Even more problematic was that the presenter chose type sizes between 10 and 12 points—far too small given that the room seated thirty. Causing even more problems was that the presenter chose a color combination of bright red lettering against a white background, a combination that would have been difficult to read even with a bold sans serif typeface, such as **Arial**, at 24 or 18 points. Worst of all, the presenter had placed by far too many words and almost no images on the slides. For thirty minutes, this engineer flipped through these presentation slides, most of the time with his body turned to the screen reading what he had created. Meanwhile, the audience listened halfheartedly and regretted that they had come.

Few slide designs used at scientific conferences and in technical meetings communicate as effectively as they should. One reason is that the defaults and templates in the most common program used for creating these slides (Microsoft's PowerPoint) do not serve scientific presentations. This section not only challenges these defaults and templates, but also proposes specific guidelines for typography, color, and layout of slides for scientific presentations.

Presented in Table 4-2 is a summary of these guidelines. Some guidelines, such as number 1 on layout and numbers 1 and 4 on style, go against the defaults of PowerPoint and therefore against what are commonly projected at conferences, meetings, and university lectures. Reasons exist for this break with tradition. The most important of these concern the differences between scientific presentations and general business presentations. First, the content of scientific presentations is typically specific and complex. Second, the audience at a scientific presentation is usually taxed to understand the content. Third, generally in a scientific presentation, images are essential for that understanding. Accompanying Table 4-2 is Figure 4-6, which gives a template for slides. An assumption for this template is that the type is sized for a presentation room that can accommodate one hundred or so. For larger rooms, the presenter might need to increase the type size.

An assumption for both Table 4-2 and Figure 4-6 is that the primary goal of the scientific presentation is to inform or to persuade an audience about technical results. In doing so, the presenter strives to have the audience remember those results after the presentation and to understand the steps for how those results were reached. Given the diversity of scientific presentations, such is not always the case. For instance, in a classroom lecture, the speaker often wants to emphasize the pro-

Table 4-2. Guidelines for slides at a scientific presentation.

Typography
Use a sans serif typeface such as Arial
Use boldface (**Arial**)
Use type sizes at least 18 points (14 points okay for references)
Avoid presenting text in all capital letters
Color
Use either light type against a dark background or dark type against a light background
Avoid red–green combinations (many people cannot distinguish)
Layout
Use a sentence headline for every slide, but the title slide; left justify the headline in the slide's upper left corner
Keep text blocks, such as headlines and listed items, to no more than two lines
Keep lists to two, three, or four items; make listed items parallel; avoid sublists, if possible
Be generous with white space
Style
Include an image on every slide
Make the mapping slide memorable; for instance, couple each section of the talk with an image that is repeated in that section
Limit the number of items on each slide
Limit the number of slides so that you can dedicate at least one minute to each

cess for solving a problem, rather than the results of the problem. Likewise, in a management presentation, the speaker sometimes wants to communicate only the results and not the steps that led to those results. For that reason, such speakers might have justification to deviate from the design guidelines advocated here. Sadly, some speakers present so many slides and pack them with so much detail that the goal seems to be neither to inform nor to persuade the audience. Rather the goal appears to be to impress the audience. For such speakers, these guidelines do not apply.

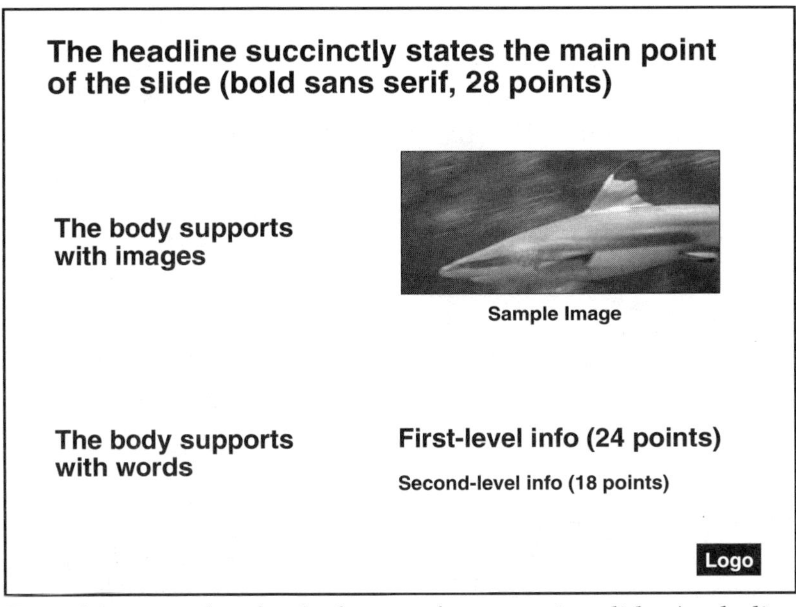

Figure 4-6. A template for the format of presentation slides (excluding the title slide). The type sizes on this template are appropriate for a room that can accommodate one hundred. For a larger room, one might need larger type sizes so that people in the back can read the text. For an example title slide, see Figure 3-8 on page 70.

Guidelines for Typography

The typography of a document, be it a journal article or presentation slide, communicates much about the document. One important choice in typography is the selection of a typestyle (also called *font*). For instance, Garamond conveys a sense of tradition in documents, which is why Garamond is used in several journals. Garamond belongs to a class of typestyles known as serif fonts, which have projecting short strokes, such as the little feet on a serif "m." Another category of typestyles is sans serif. These fonts do not have the projecting strokes (consider a sans serif "m"). One of the most common sans serif fonts is Arial. Other important choices of typography include the type size, the choice of all capitals or lowercase, and the choice of bold, italic, or normal type.

1. For presentation slides, use a sans serif font rather than a serif font. Just because a typestyle such as Book Antiqua, Times New Roman, or Garamond is appropriate for reports and papers does not mean that it is appropriate for presentation slides. The most important consideration in choosing a typestyle for a presentation slide is not tradition, but reading speed. In a presentation, reading speed is important because the audience members, splitting their concentration between what the presenter shows and what the presenter says, give themselves only a few seconds to read each projected slide. In general, when the numbers of words are few such as on presentation slides, sans serif fonts are read more quickly—especially by those looking at the screen from the sides of the room. Within this category of sans serif typestyles (see Table 4-3), you might choose Arial or Univers for formal situations and Comic Sans MS for less formal situations.

For some reason, the default typestyle of Microsoft's PowerPoint is Times New Roman, a serif font that is not read as quickly as sans serif fonts are. That difference in reading speed is especially noticeable for an audience seated on the sides of a room rather than in the room's middle. When viewed from a sharp angle, sans serif type is easier to read than serif type. That difference in reading speed is also particularly noticeable when the optics

Table 4-3. Common type faces appropriate for presentation slides.

Typestyle	Example
Arial	A body in motion will remain in motion
Arial Narrow	A body in motion will remain in motion
Comic Sans MS	A body in motion will remain in motion
Univers	A body in motion will remain in motion

of the projection system are not optimal. Such was the case when a manager recently held a ninety-minute meeting for twenty engineers off site. Because the computer projector's bulb had degraded and because the manager had chosen this default typestyle, the slides were unreadable, and the manager and attending engineers were frustrated. Conversely, using that same projector the next day, another presenter projected readable slides that relied on a bold sans serif font.

2. *For presentation slides, use boldface.* In addition to advocating a sans serif typestyle, many graphic designers also recommend using the bold version of that typestyle. Boldfacing the letters (**Arial** or **Comic Sans MS**) makes the letters more readable from a greater distance. Boldface also allows the lettering to reproduce better when placed onto an overhead transparency. Again, for some reason, the default of Microsoft's PowerPoint does not call for a boldface type.

While boldface is recommended for presentation slides, other options such as italic, underline, and outline are not. Granted, in instructional documents, italic type in small blocks is useful for emphasis. However, on presentation slides, italic type is too slow to read (particularly when viewed from the sides of the room).

3. *Choose an appropriate type size for the room.* The size of the type is also a consideration. The size of type is measured in points (a point is about 1/72 of an inch). When a bold sans serif font is used, appropriate type sizes for all slides except the title slide are between 18 and 28 points, as shown in Table 4-4 (for the title on a title slide, using 32 or 36 points is appropriate). Not surprisingly, if the default of Microsoft's PowerPoint is used (an unbolded serif font), the presenter has to use a larger type size for legibility. That is why the default type size for headlines in PowerPoint is 44 points.

Table 4-4. Recommended type sizes for presentation slides.

Size	Use
28 points	headline of slide
24 points	primary type for body of slide
18 points	secondary type for body of slide
14 points	reference listings and logos

For footnotes that the speaker does not expect the audience to actually read during the presentation (but may want to read on a copy of the presentation slides afterwards), 14 points is appropriate as long as it is clear to the audience that the text block is a footnote. Also, for a small room such as a conference room, you can drop down one size level: 24 points for the headline and 18 points for the primary text of the body. Likewise, for a large auditorium, you might consider increasing the size to 32 points (or even larger) for the headline and 28 points for the primary text of the body.

At a recent national conference in which presenter after presenter used 12 and even 10 point type on their slides—a size that people sitting in only the first couple of rows could read—one person in the audience decided that he had had enough. This person moved to the back of the auditorium, stood on a chair, and focused a pair of binoculars onto the screen. Because most of the audience members had long since given up trying to read the tiny lettering on the screen, they soon spotted the man in the back with the binoculars. A wave of laughter passed over the auditorium. The commotion was so loud that the presenter became flustered and turned off the projector. For this presenter, I have no sympathy. Not taking the time to create a slide that the entire audience can read is inconsiderate.[3]

4. Avoid presenting text in all capital letters. Many presenters mistakenly use all capital letters on their slides. These presenters fail to recognize that readers recognize words not only by the letters in the word, but also by the shape of the letters: for instance, the shapes of ascenders such as *b, d,* and *f* and the shapes of descenders such as *g, j,* and *p*. As shown in Figure 4-7, using all capital letters dramatically slows the reading because using all capitals prevents readers from recognizing the shapes of words.

Another problem with using all capitals is that type set in all capitals takes up much more space (about 35 percent more space) than type set in upper and lower case.[4] On a presentation slide, space is valuable, and what space you do not need for type and images, you want to leave blank, to make the slide more inviting to read.

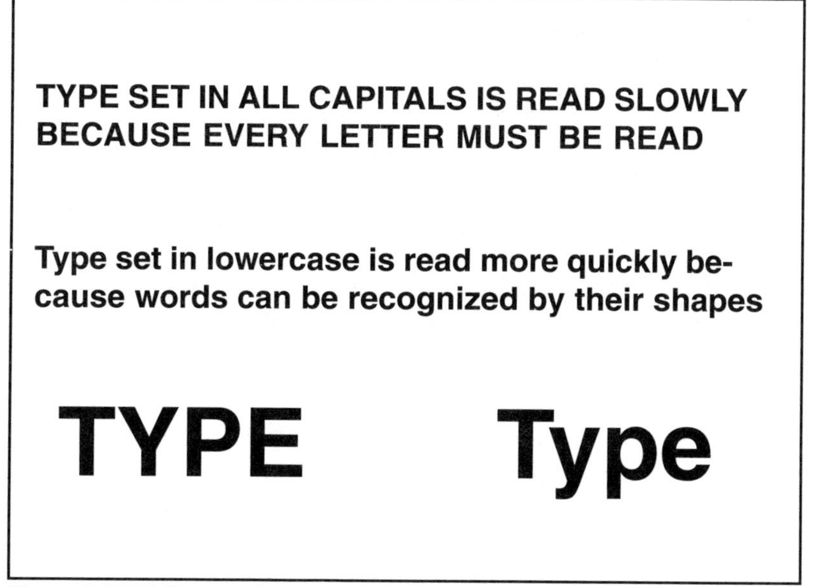

Figure 4–7. Difference between reading all capital letters and type set in uppercase and lowercase letters.

The Morton Thiokol presentation slides referred to in Chapter 2 demonstrate how using all capitals makes it difficult to discern the message. One of the slides from that presentation is shown in Figure 4-8. Not only does the slide suffer from having too many words, but all the lettering is in capital letters. The slide would have been easier to read if the presenters had used upper and lower case. For the slide to have been effective, though, the presenters would have had to use fewer words, written a sentence headline that clearly indicated the slide's main point, and incorporated an image to support that point.

Weak Slide

PRIMARY CONCERNS -

FIELD JOINT - HIGHEST CONCERN

- EROSION PENETRATION OF PRIMARY SEAL REQUIRES RELIABLE SECONDARY SEAL FOR PRESSURE INTEGRITY
  - IGNITION TRANSIENT - (0-600 MS)
    - (0-170 MS) HIGH PROBABILITY OF RELIABLE SECONDARY SEAL
    - (170-330 MS) REDUCED PROBABILITY OF RELIABLE SECONDARY SEAL
    - (330-600 MS) HIGH PROBABILITY OF NO SECONDARY SEAL CAPABILITY

- STEADY STATE - (600 MS - 2 MINUTES)
  - IF EROSION PENETRATES PRIMARY O-RING SEAL - HIGH PROBABILITY OF NO SECONDARY SEAL CAPABILITY
    - BENCH TESTING SHOWED O-RING NOT CAPABLE OF MAINTAINING CONTACT WITH METAL PARTS GAP OPERATING TO MEOP
    - BENCH TESTING SHOWED CAPABILITY TO MAINTAIN O-RING CONTACT DURING INITIAL PHASE (0 - 170 MS) OF TRANSIENT

Figure 4–8. Weak presentation slide from Morton Thiokol presentation to NASA on January 27, 1986.[5] The use of all capital letters makes reading difficult. Also a problem is having so many words on the slide, far too many words for the audience to comprehend in a presentation.

Guidelines for Color

At a thirty-minute contractors' presentation before sixty engineers and scientists, most of whom had flown to the meeting, an engineer projected a set of computer slides with a dark rust type against a brown background. The audience at first thought that the engineer had begun with a blank brown background upon which he would build a slide. Unfortunately, nothing ever appeared. The engineer proceeded to project another slide with the same blank brown background, and the audience began whispering among themselves. The speaker, sensing the agitation from his audience, turned to look at the screen. Not even he, standing a couple feet from the screen, could read the words. This engineer, who later claimed that he could see the contrast on his computer screen, had neglected to try out the color combination on a projected screen. Even if the engineer could see the contrast on his computer screen, he should have given more thought to the colors that he had chosen.

1. Consider the representative colors of your institution. For a company that has blue as its identifying color, incorporating blue into the color scheme of its presentation slides is natural. Sandia National Laboratories, for instance, uses blue as an identifying color. For that reason, many presentation slides representing Sandia use blue—either blue lettering on a white background or white lettering on a blue background. Likewise, Lawrence Livermore National Laboratory has green as an identifying color and uses green in a similar fashion.

The color associated with your institution should not be the only consideration in choosing a color. For instance, I teach at Virginia Tech, which has maroon and orange as its representative colors. While these colors work well on our team's football jerseys amid the beautiful autumn leaves of the Appalachian mountains, these colors do not work well as the main colors on presentation slides. For that reason, I use maroon as an accent color on the slides, but choose a cool color combination, such as white lettering on a dark blue background, as the main color combination for the slide. When printing out overhead transparencies or handout pages, I usually reverse that combination (blue lettering against a clear background) to save on toner.

2. *Consider how readable the combination is.* As you might infer from the anecdote about the engineer who used rust letters against a brown background, choosing a color combination with a high contrast is important. Not all color combinations are read with equal speed. The color combination that is read most quickly is black lettering against a yellow background,[6] which is one reason that caution signs use this combination. The next most quickly read combination is black lettering against a white background. One of the slowest-to-read combinations is black lettering against a red background, and even more slowly read is red lettering against a black background. Although dark blue or dark green lettering against a white background is not read as quickly as black against a white or yellow background, these combinations can be read quickly enough to serve a scientific presentation. In the end, what is important is that the contrast be high.

Another consideration is color blindness. About 8 percent of males and 0.5 percent of females have deficiencies in distinguishing certain color combinations.[7] The combinations that cause the most problems for these people involve red, green, and brown. For that reason, avoid such combinations.

3. *Consider the effect of the background color upon the audience.* Blue and green are soothing colors. For that reason, audiences feel comfortable with either of those colors used as the background of a slide. Orange and red, on the other hand, are hot colors and can unsettle an audience. Unless you desire to rile an audience, avoid such colors as your background color. Even yellow as a background color can agitate. When I first learned that black against yellow was the fastest color combination to read, I tried that combination on my overhead transparencies.

That semester the students seemed unusually agitated in class, and their questions were often caustic. When I switched to white lettering against a blue background, the students calmed down noticeably. Interestingly, yellow lettering against a blue background does not have nearly the same agitating effect—unless you have too much yellow text and too many yellow lines.

Guidelines for Layout

On presentation slides, one of the main layout errors is having too many details. When a slide has too many details, the listeners are intimidated; they feel that they do not have time both to decipher the slide and to continue listening to the speaker. Specifically, what intimidates audiences are slides with large blocks of text (more than two lines per block), slides with long lists (more than four items per list), and slides that do not contain enough white space.

A second layout error that causes slides not to communicate effectively is an illogical arrangement of information. When a slide is projected, the audience turns from the speaker and looks at the screen. At this point, the audience's attention is divided between the speaker and the slide. For this situation, it is important that the audience members quickly grasp the purpose of the slide and that they know how to read it: what to read first, what to read second, and so on. In poorly designed slides, the audience does not know on what to focus first.

Given in this section are guidelines for limiting the amount of information so that the audience is not overwhelmed. Also given here are guidelines for arranging the information so that the audience is quickly oriented.

1. *For all slides except for the title slide, use a sentence headline to state the slide's purpose.* When you place a presentation slide before the audience, the audience immediately

turns to it and tries to decipher its purpose. A sentence headline, such as shown on the slide of Figure 4-9, serves this situation by orienting the audience quickly to the purpose of the slide so that the audience can turn its attention back to the presenter. Designing slides with short sentence headlines is not a new idea. For instance, in the 1960s, Robert Perry at Hughes Aircraft began advocating sentence headlines for slides, and since the 1970s, Larry Gottlieb of Lawrence Livermore National Laboratory has taught the design to hundreds of scientists and engineers.[8]

For a sentence headline to be effective, you should follow three principles. First, the sentence headline should begin in the upper-left corner of the slide. That way, the audience sees it first. Second, the sentence headline should be no more than two lines. Blocks of text longer than two lines on a slide are often not read. Third,

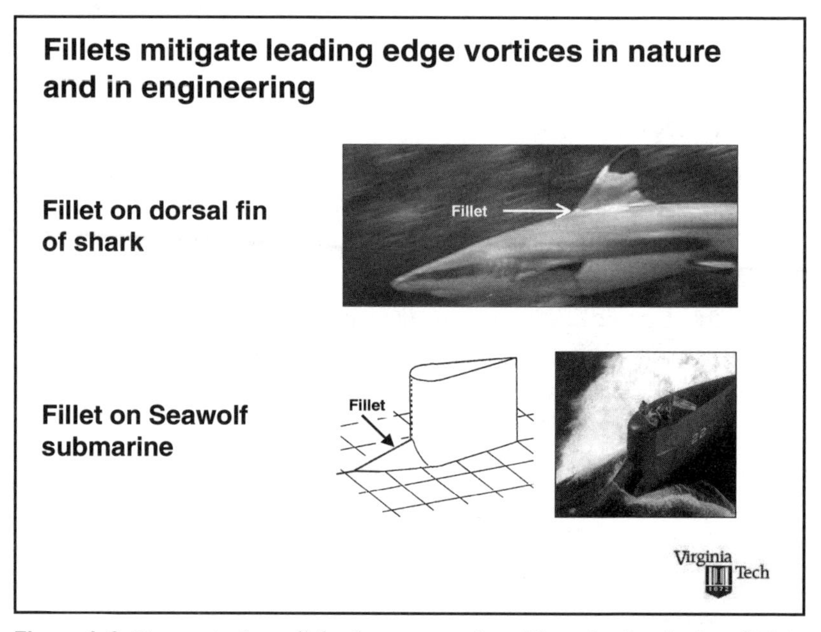

Figure 4–9. Presentation slide that uses a headline–body design.[9] This slide comes from a presentation that introduced a fillet design for reducing the vortices that occur along turbine vanes in a gas turbine engine.

to make it easier for the audience to read, the headline should be left justified, rather than centered, because a centered headline takes the audience longer to read, particularly if the headline goes to a second line.

Using a sentence headline is not the norm in scientific presentations. In fact, given the thousands of presentations that use phrase headlines (or, worse yet, no headlines), this advice swims against the current of what is most often seen. However, good reasons exist for using sentence headlines.

The first reason is that while a phrase headline identifies the topic, a sentence headline can show a specific perspective on the topic. Contrast the phrase headline in the top slide of Figure 4-10 with the sentence headline of the bottom slide. The sentence headline in the bottom slide orients the audience much more effectively.

A second reason to use sentence headlines is that although the speaker might make smooth transitions between slides, the audience might not catch those transitions. Often in presentations, a topic on one slide will cause members of the audience to think about their own work for a moment, which can cause them to miss the speaker's transition to the next slide. The sentence headline allows those audience members to reorient themselves. Also, for those situations in which the speaker distributes or posts the presentation slides as a handout, the sentence headlines have many advantages over phrase headlines. For instance, an audience that views the slides weeks later or an audience that was not able to attend the presentation is in a much better position to see the organization, emphasis, and transitions of the talk if sentence headlines are used. In fact, when the sentence headlines are well written, the slides can serve as an informal report for the work.

A third reason to use sentence headlines is that a sentence headline not only orients the listener more effectively, but also orients the speaker more effectively

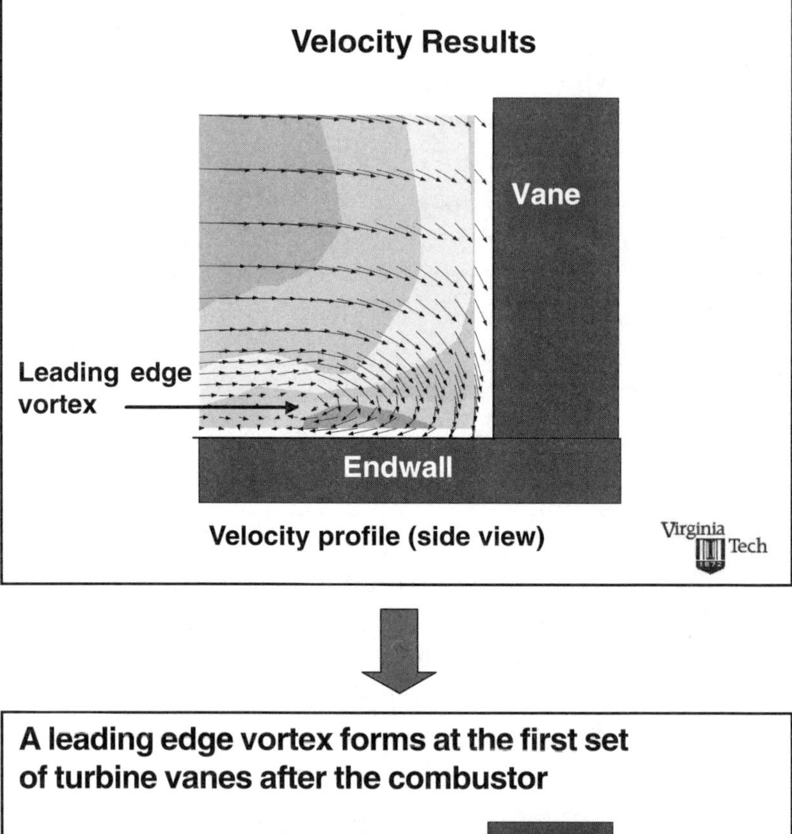

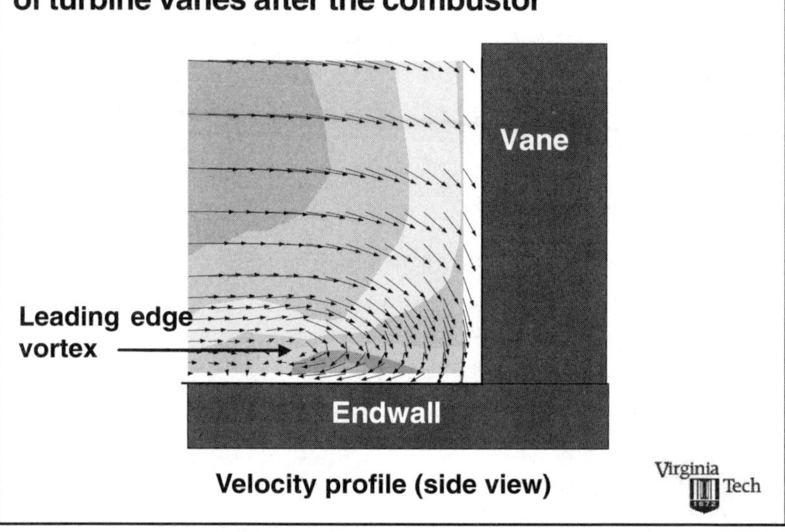

Figure 4-10. Two slides: (top) weaker slide with phrase headline, and (bottom) stronger slide with sentence headline.[10]

during the presentation. Often in science and engineering, people must make presentations that they did not themselves create. By having sentence headlines, the speaker can see what main point each slide has and can make the appropriate transition to that point.

Yet a fourth reason, and what my colleague Harry Robertshaw considers the most important, is that a sentence headline forces the presenter to come to grips with what the main purpose of the slide is. This point might seem obvious, but in the presentation slides sent by Morton Thiokol to NASA the night before the launch of the space shuttle *Challenger*,[11] the presenters did not make clear their assertions and did not provide enough evidence for the assertions that they did make. Consider the second slide of that presentation again (Figure 4-11). If the presenting engineers had simply stated their main assertion "The lower the temperature of the launch, the more erosion the O-rings have experienced," then the engineers might have reconsidered the evidence that they presented. Perhaps they would have come up with a graph, like that suggested by Edward Tufte in *Visual Explanations*,[12] that plotted an erosion index for the O-rings versus launch temperature for all launches up to that point. As it was, Morton Thiokol had no real headline on that slide, and the evidence in the slide's body lacked the key relationship between O-ring erosion and launch temperature.

2. *Achieve a balance between what you say and what you show.* Once you have established the purpose of the slide with the sentence headline, you should support that assertion with images and words in the slide's body. In general, you want enough images and words to support the headline's assertion, but not so many that the audience is overwhelmed. One example of a slide having a nice balance of white space with supporting words and images was shown in Figure 4-9.

HISTORY OF O-RING DAMAGE ON SRM FIELD JOINTS

	SRM No.	Cross Sectional View: Erosion Depth (in.)	Cross Sectional View: Perimeter Affected (deg)	Cross Sectional View: Nominal Dia. (in.)	Top View: Length Of Max Erosion (in.)	Top View: Total Heat Affected Length (in.)	Clocking Location (deg)
61A LH Center Field**	22A	None	None	0.280	None	None	36° - 66°
61A LH CENTER FIELD**	22A	NONE	NONE	0.280	NONE	NONE	338° - 18°
51C LH Forward Field**	15A	0.010	154.0	0.280	4.25	5.25	163
51C RH Center Field (prim)***	15B	0.038	130.0	0.280	12.50	58.75	354
51C RH Center Field (sec)***	15B	None	45.0	0.280	None	29.50	354
410 RH Forward Field	13B	0.028	110.0	0.280	3.00	None	275
41C LH Aft Field*	11A	None	None	0.280	None	None	--
410 LH Forward Field	10A	0.040	217.0	0.280	3.00	14.50	351
STS-2 RH Aft Field	28	0.053	116.0	0.280	--	--	50

*Hot gas path detected in putty. Indication of heat on O-ring, but no damage.
**Soot behind primary O-ring.
***Soot behind primary O-ring, heat affected secondary O-ring.

Clocking rotation of leak check port - 0 deg.

OTHER SRM-15 FIELD JOINTS HAD NO BLOWHOLES IN PUTTY AND NO SOOT HEAR OR BEYOND THE PRIMARY O-RING

SRM-22 FORWARD FIELD JOINT HAD PUTTY PATH TO PRIMARY 0-RING, BUT NO O-RING EROSION AND NO SOOT BLOWBY. OTHER SRM-22 FIELD JOINTS HAD NO BLOWHOLES IN PUTTY.

Figure 4-11. Weak slide that Morton Thiokol sent to NASA to request launch delay of the space shuttle *Challenger* (January 27, 1986).[13] If the presenters had chosen an effective headline (such as "The lower the temperature of the launch, the more erosion that the O-rings experienced"), perhaps they would have seen that the key relationship of O-ring erosion to launch temperature was missing from the slide.

A common mistake in designing the bodies of slides is not achieving a balance between the words that are said and the words that are shown. In a strong presentation, although you often repeat words and phrases from the slides in your speech, your speech should include more than just the words on the slides—much more. In many weak presentations, all the words that the speaker says are given on the slides. Consider the slide shown in Figure 4-12, which my wife recently witnessed.[14] This slide and similar ones in that presentation irritated the audience. The audience was not sure whether to listen or to read. In the end, they did neither. When the presenter asked for questions at the end of the presentation, a long and uncomfortable silence ensued. That silence contrasted sharply with all the questions and discussion of

Weak Slide

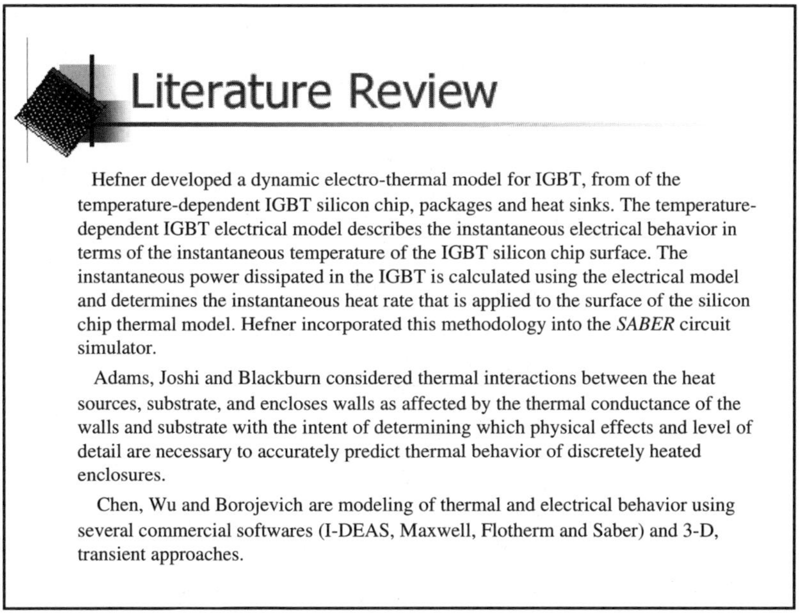

Literature Review

Hefner developed a dynamic electro-thermal model for IGBT, from of the temperature-dependent IGBT silicon chip, packages and heat sinks. The temperature-dependent IGBT electrical model describes the instantaneous electrical behavior in terms of the instantaneous temperature of the IGBT silicon chip surface. The instantaneous power dissipated in the IGBT is calculated using the electrical model and determines the instantaneous heat rate that is applied to the surface of the silicon chip thermal model. Hefner incorporated this methodology into the *SABER* circuit simulator.

Adams, Joshi and Blackburn considered thermal interactions between the heat sources, substrate, and encloses walls as affected by the thermal conductance of the walls and substrate with the intent of determining which physical effects and level of detail are necessary to accurately predict thermal behavior of discretely heated enclosures.

Chen, Wu and Borojevich are modeling of thermal and electrical behavior using several commercial softwares (I-DEAS, Maxwell, Flotherm and Saber) and 3-D, transient approaches.

Figure 4-12. Slide in which the presenter placed the entire speech onto the slide. Not surprisingly, no one in the audience bothered to read this slide.

the other presentations that morning. This presenter's work was not only not well received; it was not received at all.

Another story that points to the importance of having a balance between what you say in your speech and what you show on your slides occurred several years ago at one of the national laboratories.[15] On this occasion, the Secretary of Energy was visiting and attended a presentation given by a department manager. The department manager had worked for weeks on this presentation. He had booked the best conference room at the lab, he had recruited the best artists at the lab to design the slides, and he had practiced the presentation over and over until he could say every word on the slides without even looking at the slides. After the third slide, though, the Secretary of Energy raised his hand. The department manager stopped and said, "Yes, you have a question?"

"No. No, I don't have a question," the Secretary of

Energy said. "I have a comment. I can read. From now on, don't say anything else. Just put the slides up one by one. I'll tell you when to change them."

As you might imagine, the department manager was humiliated. For this department manager, I have sympathy. Given the work that he put into his presentation, he deserved better treatment.

How much wording should be placed onto slides? My rule of thumb is to keep each block of text, including the headline, to no more than two lines. Audiences are much more likely to read blocks of text with one or two lines than blocks that are longer.

3. Avoid lists with more than four items. Genesis, Exodus, Leviticus, Numbers, Deuteronomy, First and Second Samuel, First and Second Kings—that list continues for another fifty-seven items. While we sometimes spend hours memorizing long lists such as the books of the Bible, we expect too much of our audiences when we ask them to remember long lists that we display for only a minute or so in our presentations. As mentioned in Critical Error 4, audiences remember lists of twos, threes, and fours. In a presentation, lists that have more items are soon forgotten. Worse yet, the audience often does not even try to read long lists. With a long list, the audience sees the length, perhaps reads the first couple of items, and then turns away. Presenters would do better to place only the four most important items from the list on the slide and reserve the less important details for the speech.

What if your work contains a set larger than four? For instance, what if you are evaluating seven characteristics of a receiver at a solar energy plant:

Steady-state efficiency
Average efficiency
Startup time
Operation time
Operation during cloud transients
Panel mechanical supports
Tube leaks

Rather than giving your audience all seven characteristics up front, consider placing the characteristics into more memorable groups. One example is as follows:

Efficiency of receiver
Operation cycle of receiver
Mechanical wear on receiver

When you discuss the efficiency of the receiver, you can then introduce steady-state efficiency and average efficiency as a group of two characteristics for that category. Likewise, when you introduce mechanical wear on the receiver, you can introduce panel mechanical supports and tube leaks as another group of two characteristics for that category. The advantage is that the audience is much more likely to recall the list of three categories than the longer list of seven characteristics.

An exception to excluding a long list is the case in which the presenter does not expect the audience to actually read the list. Rather, the presenter just wants the audience to see that many examples exist. For instance, a presenter might want to show the many negative effects of a drug. In this case, the presenter might use a long list of examples as overwhelming evidence for the assertion that this drug is dangerous.

No matter what the purpose of the list is, the items in the list should be parallel in structure. In other words, if the first item of a list is a noun phrase, all items should be noun phrases. In addition, if you include one subitem, logic dictates that you include a second. If possible, avoid sublists because audiences usually do not read them. That sublevel of information is better left in your speech.

4. Avoid unnecessary details. As mentioned, once you have established the purpose of the slide with the sentence headline, you should support that assertion with words and images in the body of the slide. A common error, though, is to place too many supporting details onto each slide. Placing all the details of your work on the presentation slides causes the audience to lose sight of the details that are most important. In other words, by placing too many details on your presentation slides, you run the risk of the audience not remembering the most important details. Worse yet, in cases such as that shown in Figure 4-13, you risk having the audience give up without even trying to understand the slide.

One way to prevent a slide from seeming overcrowded is to limit the number of items on the slide. Many graphic designers recommend a maximum of seven items. Figure 4-14 provides an example. This slide has seven main parts: the headline, the image, the three callouts, the sentence in the body, and the logo. What makes this slide readable is the white space that allows the audience to separate these items. This white space also allows the audience to find an order in which to read the information: in this case top to bottom. Contrast that order with the lack of order in Figure 4-13.

Weak Slide

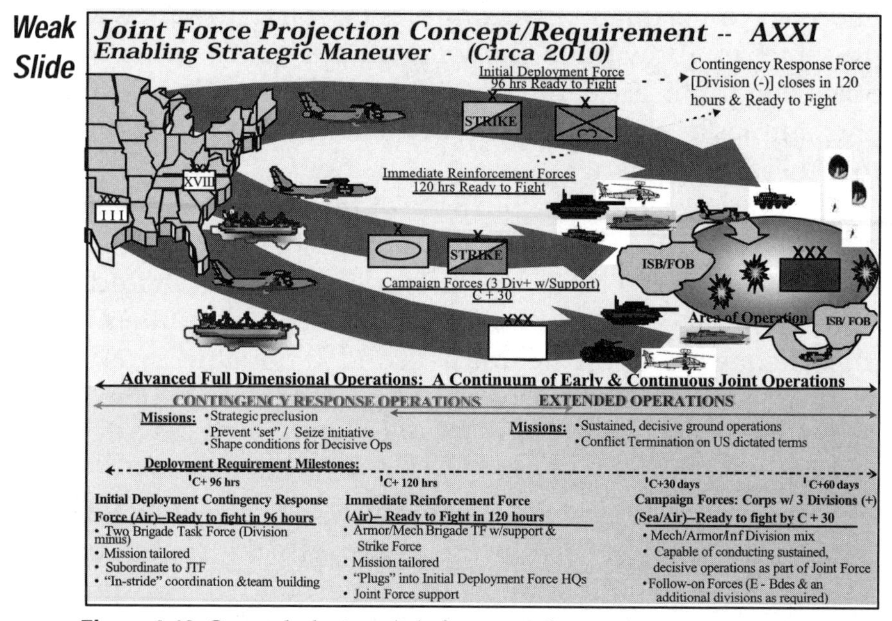

Figure 4-13. Overwhelming slide from a military presentation. Although the presenter put much effort into making this slide, this slide overwhelms because there are too many details.

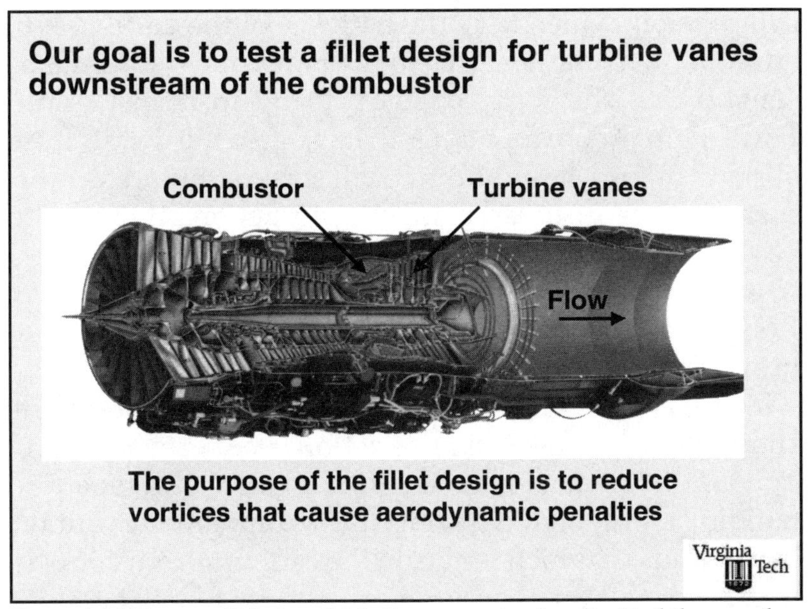

Figure 4-14. Strong slide in which the presenter has limited the number of details and arranged those details to allow enough white space.[16]

What if you have more than seven details to convey to the audience? How would you work those into the presentation? One way, if time allows, would be to have a second slide. Another way would be to present the secondary details in the speech. Granted, the audience will not be as likely to remember the secondary results if they are placed in the speech, but if the speaker packs every result and image into his or her presentation slides, the audience is likely not to remember any details, not even the primary ones.

A third way to work in more than seven items is to add them during the presentation. In a computer projection, this adding (or building) is easy: You have the program bring in additional items after the audience has digested the ones you have shown. With an overhead transparency, you can achieve the same effect by using overlays.

When building a slide, be careful about having too many stages. Some presenters go overboard and build every detail, which tests the patience of the audience. In addition to being sensitive to the amount of building, be sensitive to the way that you bring in items. Avoid PowerPoint's cute functions that bring in the details from all sorts of directions and with all sorts of fanfare. Unfortunately, one of those distracting functions happens to be PowerPoint's default (*Fly from left*), which calls for items to stream in from the left. As my colleague Harry Robertshaw points out, a much less distracting way to bring items on the screen is the choice named *Appear*, which has the item simply and quickly appear on the slide. Although the *Appear* selection is not easy to find in PowerPoint, it is worth the effort. Finally, with regard to building a slide, avoid having any accompanying sounds. These sounds, which range on PowerPoint from clicks to whooshes to brakes screeching, just grate on the audience and have no place in a professional presentation.

Besides having too many details, many slides in scientific presentations suffer because the details contain too much complex mathematics. It is unreasonable to expect your audience to follow complex mathematics when you do not have the time to methodically work through that mathematics. I am not saying that you should remove all complex equations from the slides of a short presentation. What I am recommending is that when you show mathematics, you account for what the audience can comprehend during the presentation. If the presentation allots the audience enough time to follow your entire derivation, so be it. However, if the audience does not have the time to follow the derivation, then you should clarify for them what you expect them to gather from the display of the mathematics.

For instance, in showing a complex equation, you could state up front that you do not expect the audience to follow all the mathematics. Rather, you have shown this equation to point out what the terms physically represent. For instance, the first term might represent the rate of mass flow out of the control volume, the second term might represent the rate of mass flow into the control volume, and so on. By clarifying what you expect the audience to gather, you allow them to relax. Without that clarification, though, some in your audience will simply quit listening to the presentation because they realize that they have no hope of working through the mathematics.

Other slides suffer because the illustrations are too complex for the audience to absorb. For instance, the illustration on the slide in Figure 4-15 is much too detailed for an audience to digest in two minutes. In such situations, the presenter has to decide which details are important for the audience to understand. For example, if all the information in Figure 4-15 has to be communicated to the audience, then the slide should be split into two, possibly three, slides, with one slide focusing on the direction of the mission and another focusing on the timeline. In regard to the timeline, if all the details are important, so be it. However, if some are secondary, consider showing them in a muted way (perhaps in a light gray), so that the key details stand out and the audience is not overwhelmed by the graphic.

This chapter has challenged several defaults of Microsoft's PowerPoint. A summary of these challenges can be found in Table 4-5. In addition to the challenges already discussed, two other challenges arise on the grounds that these defaults (or templates) create unnecessary details. One challenge is to the background designs that PowerPoint makes available as templates. Fireballs, meadow scenes, ribbons, party balloons—these backgrounds might be appropriate for fund-raising presentations at a fraternity house, but are distractions in scientific presentations. A much better choice of background is a dark blue or green with white or yellow for the type. Another good choice for the background is a very light color with a dark color for the type. To make a background color distinctive, the airbrush option on

Weak Slide

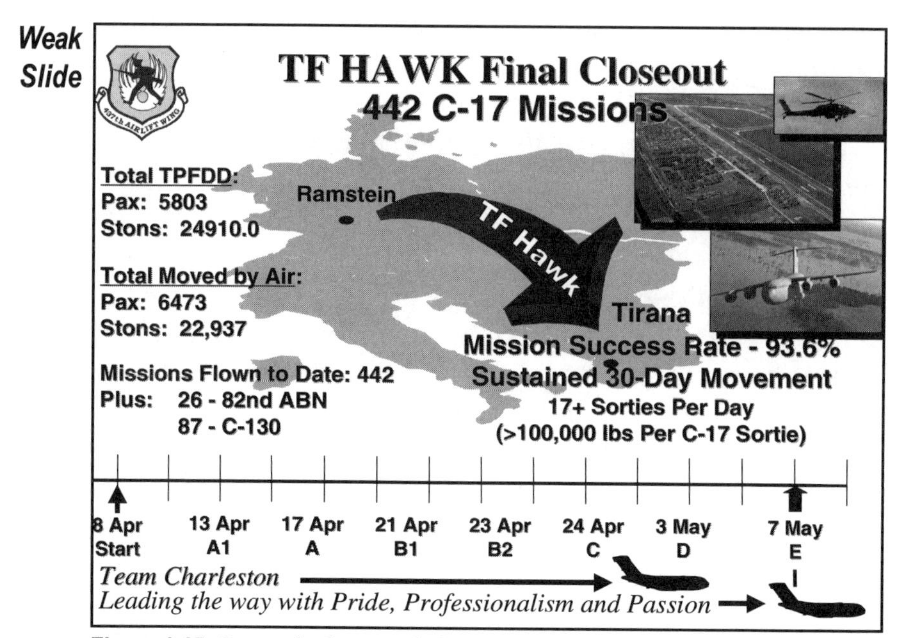

Figure 4-15. Overwhelming slide. A possible revision would break up the slide into two slides: one with the map and one with the timeline.

Table 4-5. Format defaults in Microsoft's PowerPoint that should be challenged for slides in scientific presentations.

Format	PowerPoint Default	Suggested Change
Typeface	Times New Roman	**Arial Boldface**
Type in headline	Centered 44 points	Left-justified 28 points
Type size in body	32 points	24–18 points
Separation indicator		
Main item in list	Bullet	Vertical white space
Secondary item in list	Sub-bullet	Indent
Entry animation	Fly from left	Appear
Background	Various templates	Light color (dark typeface) Dark color (light typeface)

PowerPoint works well. Another factor in choosing the background color is the kind of projection to be used: overhead projection or computer projection. When printing out the slides onto transparencies or handout pages, a light-colored background is preferable to save toner on your printer. A light-colored background is also preferred if you are incorporating line graphs and line drawings from programs that create those graphs or drawings on white backgrounds.

Another challenge to the defaults of PowerPoint concerns its overuse of bullets (which are black dots to indicate a new item in a list). The main problem with bullets is that they often pull emphasis away from the words in the list and place that emphasis onto the dots. Richard Feynman did not think much of the practice of using bullets,[17] and neither do I. A much less distracting way to indicate the separation of items in a list is with extra white space placed vertically between the items of the list. Unfortunately, the defaults of PowerPoint not only call for bullets on all main text blocks, but also call for sub-bullets on any subordinate text blocks. Note that indenting subordinate points achieves the same goal without the distraction.

The overall message here is not that you should avoid programs such as Microsoft's PowerPoint. The message is that you should assess the defaults of such programs to determine whether those defaults serve your audiences, purposes, and occasions. In those cases where the program's defaults do not serve the presentations, then you should be proactive and change them.

Critical Error 6
Projecting Slides That No One Remembers

> *Approval for our 1.2 million dollar proposal came down to a short presentation with a maximum of two slides. Talk about pressure. The worst part was that I would not be making the presentation – a manager in the sponsoring program would be, and essentially all he knew about the project was the information on those two slides.*[1]
>
> —Daniel Inman

In a presentation, the audience remembers on average about 10 percent of what is said and 20 percent of what they read on projected slides. However, when the presenter both says details and shows those details on well-designed slides, the retention by the audience can climb to about 50 percent.[2] How close to 50 percent this retention reaches depends on how well the slides are designed. While the discussion for Critical Error 5 centered on how to format slides so that the retention level is high, the discussion of this critical error centers on what to place on slides so that the audience retains what is most important to remember. As mentioned, if a presenter tries to place all the details of the work onto the slides, then the presenter overwhelms the audience, and the audience ends up retaining little. For that reason, presenters have to be selective about what they include. Unfortunately, many presenters place relatively unimportant information onto slides and, in so doing, leave off details that the audience actually needs.

So what information should you include? The answer lies in the reasons for projecting slides in the first place. One important reason to include slides is to show images that are too complicated to explain with words. A second important reason is to emphasize key results. Given these two reasons, it is easy to see that slides should include the most important images and results of a presentation. Yet a third reason to include slides is to reveal the organization of the presentation. By making the audience aware of the presentation's organization, the presenter keeps the audience more relaxed because the audience knows where they are in the presentation. Since they are not worried about where they are, they are able to focus more on what the presenter communicates.

Showing Key Images

Before the shot clock became part of college basketball, some teams would try to slow games down by having the players continue to dribble and pass until they had a sure basket. In these games, the opposing crowd would often chant, "Boring, boring, boring." Boring—that describes the slides created by many scientists and engineers in a scientific presentation. In such presentations, the presenter has a stack of slides, each with a cryptic phrase headline and then a laundry list of bullets and sub-bullets. The effect of such a presentation on the audience is hypnotic—much like the repetitious swing of a hypnotist's watch.

Images are one way to make slides engaging. Moreover, because many images are difficult to communicate with only speech, you should take advantage of the opportunity that a presentation provides to display the key images of your work. The brain processes visual information much more quickly than text—400,000 times more quickly according to some researchers.[3] For a presentation on the dwindling numbers of Siberian tigers, images to include might be a photograph of a tiger in the wild, a map showing the range of tigers fifty years ago as opposed to today, and a bar chart showing the decrease in numbers over the past one hundred years. In situations for which you cannot think of an image, you should consider having at least a table with words and numbers as opposed to just a list of phrases, because the table would show the relationships of those words and numbers.

Another reason to include images is that the audience will remember images much longer than they will remember words. Think about your earliest childhood memories. Rather than words that people spoke to you, you are much more likely to remember images: white shirts hanging on a line, a neighbor's Dalmation lying in the grass, a tire swing tied to an apple tree. Likewise, when the audience tries to remember a presentation, the images that you have projected are much more likely to be recalled. Consider the difference between the top and bottom mapping slides in Figure 4-16. Although the top slide has many more words, this slide communicates much less than the bottom slide does. Note that most of the words in the body of the top slide are unnecessary. For instance, every presentation has an *Introduction* and *Conclusion*. Moreover, the word *Background* does not give enough information to help the audience. In addition, the audience should already know whether *Questions* are to occur at the end. The most important words on this slide are the words indicating what will occur in the middle of the presentation. Unfortunately, in this top slide, as in so many other mapping slides for presentations, these words are not memorable. The bottom slide, however, makes those words memorable by anchoring them with images. These images are much more likely to be recalled by the audience throughout the presentation, especially if the images are repeated at the beginning of the corresponding sections (as they were in this presentation of a fillet design for turbine vanes).

The mapping slide is not the only slide that benefits from images. All slides, including the title slide and

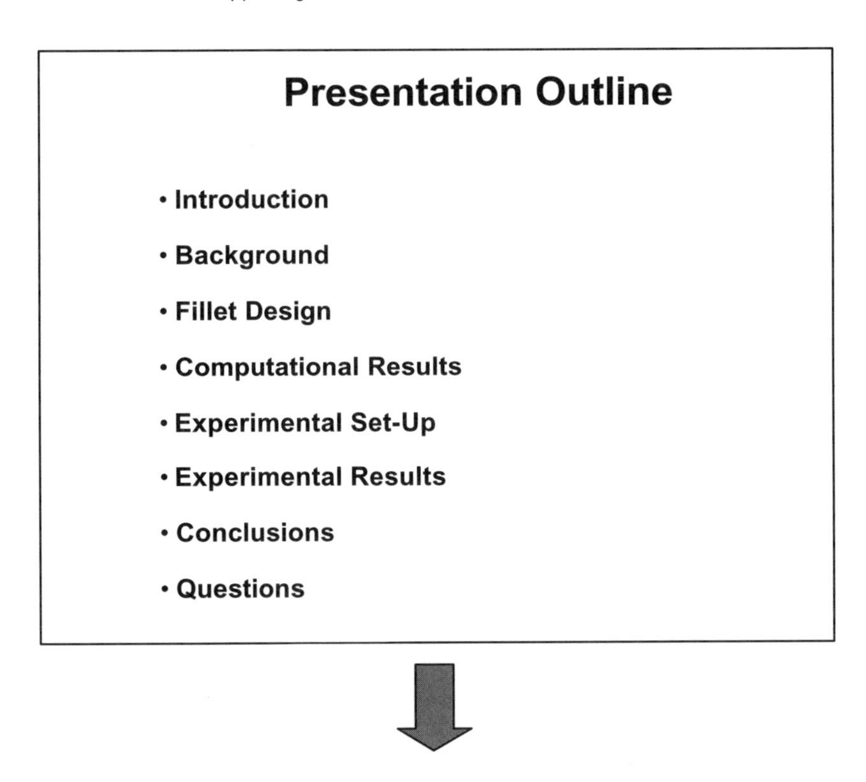

This talk presents a computational and experimental analysis of the fillet design

1. Fillet Design

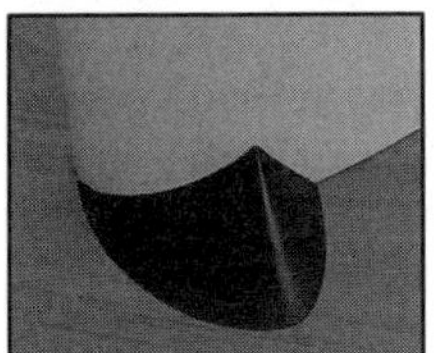

2. Computational Predictions

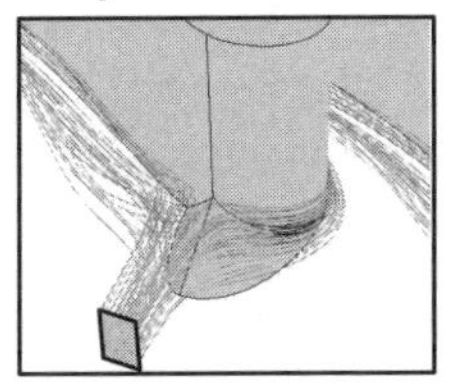

3. Experimental Set-Up

4. Experimental Results

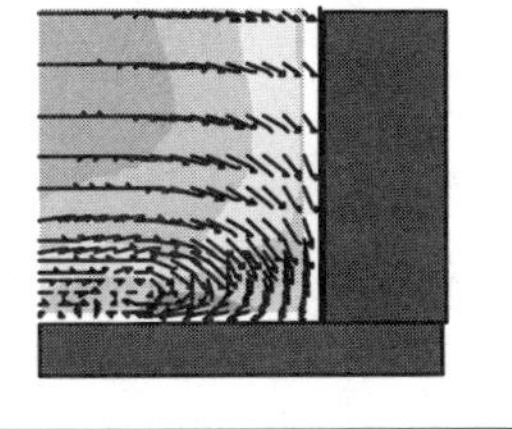

Virginia Tech

Figure 4-16. Two slides that map the same presentation: (top) weaker slide that relies solely on words, and (bottom) much more memorable slide that uses images.[4]

conclusion slide, become more memorable when a key image or icon is included. Example slides from the middle portion of the fillet presentation are presented in Figure 4-17 and Figure 4-18. Figure 4-17 shows the shape of a fillet for a turbine vane, and Figure 4-18 shows a design for a wind tunnel experiment used in testing the effectiveness of those fillet designs on preventing leading edge and horseshoe vortices.

Showing Key Results

Besides showing the presentation's key images, slides should show key results. If what you say and show has the potential of increasing recall for the audience to 50 percent, you certainly want to place the most important results of the presentation on your slides. For instance, Figure 4-19 shows computational results predicting that a fillet will prevent the horseshoe vortex and delay the passage vortex. Another example appears in Figure 4-20, which presents experimental evidence that a fillet design prevents a leading edge vortex from forming at the juncture of the turbine vane and the endwall. In showing each of these slides with a computer projector, the presenter could begin by showing the slide without the image on the right. Then the presenter could bring in that image once the audience was oriented to the image on the left.

Showing the Presentation's Organization

Besides presenting the key images and the key results on your slides, you should use slides to show the presentation's organization. By showing the organization of the presentation, you make it easier for your audience to understand how details on the slides fit into the big

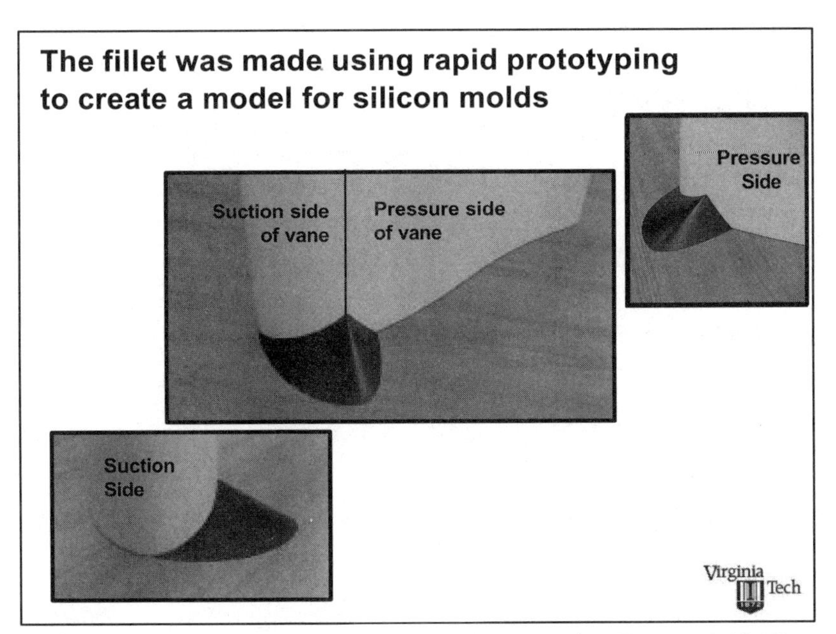

Figure 4-17. Slide from the presentation mapped in Figure 4-16. This slide shows a fillet design for a vane in a gas turbine engine.[5] For this presentation, this fillet is a key image.

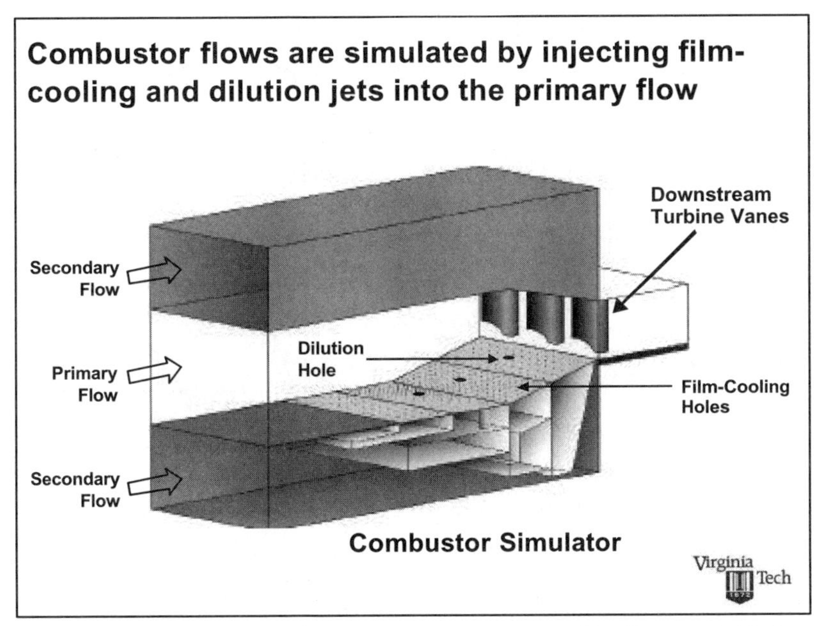

Figure 4-18. Slide from the presentation mapped in Figure 4-16 that shows the experimental setup for testing designs of gas turbine vanes.[6] This setup is another key image for the presentation.

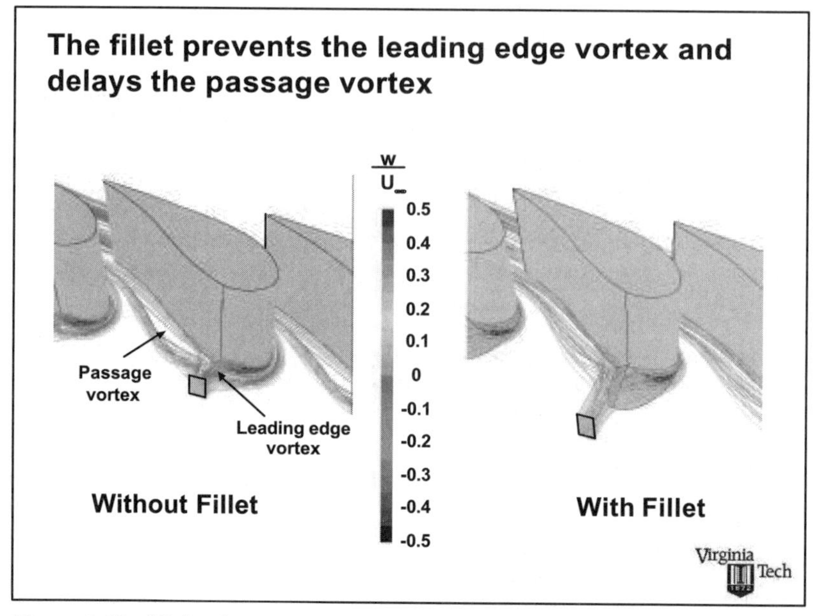

Figure 4-19. Slide showing computational predictions that a fillet prevents the leading edge vortex and delays the passage vortex.[7]

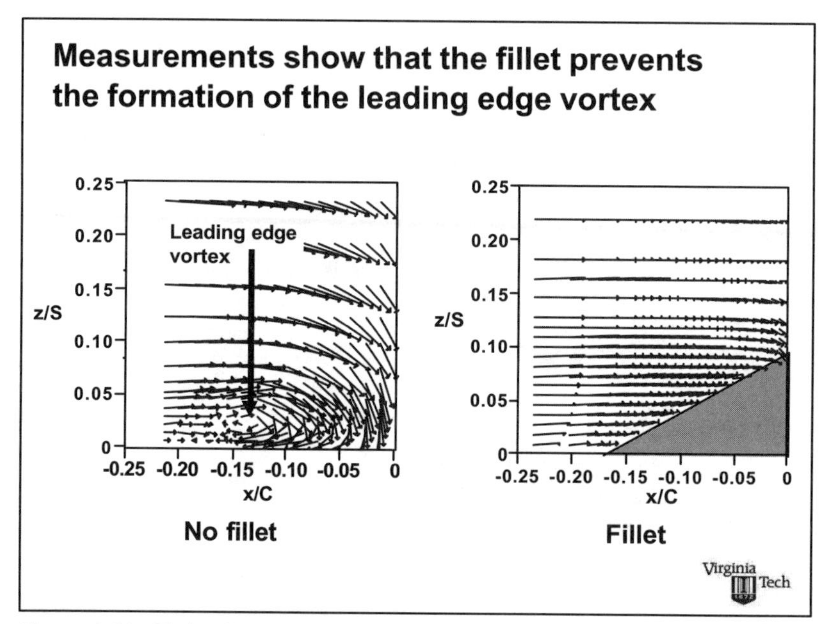

Figure 4-20. Slide showing experimental evidence that a fillet prevents the leading edge vortex.[8] In this presentation, experimental evidence showing that the fillet delays the passage vortex appears on another slide.

picture of the presentation. With that understanding, the audience can catalogue information more easily and then recall that information more readily.

In a well-designed set of presentation slides, several slides serve to reveal the presentation's organization:

Title slide
Mapping slide (showing sections of middle)
First slide for each section of middle
Conclusion slide

Figure 4-21 presents a set of these organizational slides for a presentation on ways to reduce sulfur dioxide emissions from coal-fired power plants.

The title slide contains key information: the title of the presentation in large lettering, the speaker's name and affiliation, a key image from the work, and an icon for the affiliation. Contrast this slide with the title slide (shown in Figure 2-3) that Morton Thiokol sent to NASA in their failed attempt to postpone the launch of the space shuttle *Challenger* on January 28, 1986. Morton Thiokol's slide contained a title, but that title did not reflect the ultimate purpose of the presentation: to delay the launch. Moreover, Morton Thiokol's name, the names of the engineers petitioning for the delay, and Morton Thiokol's logo did not appear on the slide. For that reason, this title slide did not carry the authority that it should have.

Another key slide that reveals the organization of a talk is the mapping slide of the presentation. In the sample presentation of Figure 4-21, this slide introduces the categories of methods that will be discussed in the presentation. Unlike typical mapping slides, this mapping slide does much more than just list the three categories of methods. This slide also depicts the process for bringing the coal to the plant, burning the coal, and emitting the combustion gases. These images provide the speaker with many opportunities to work in background information and therefore to leave this key organization slide up for a longer period of time. So often in typical presentations,

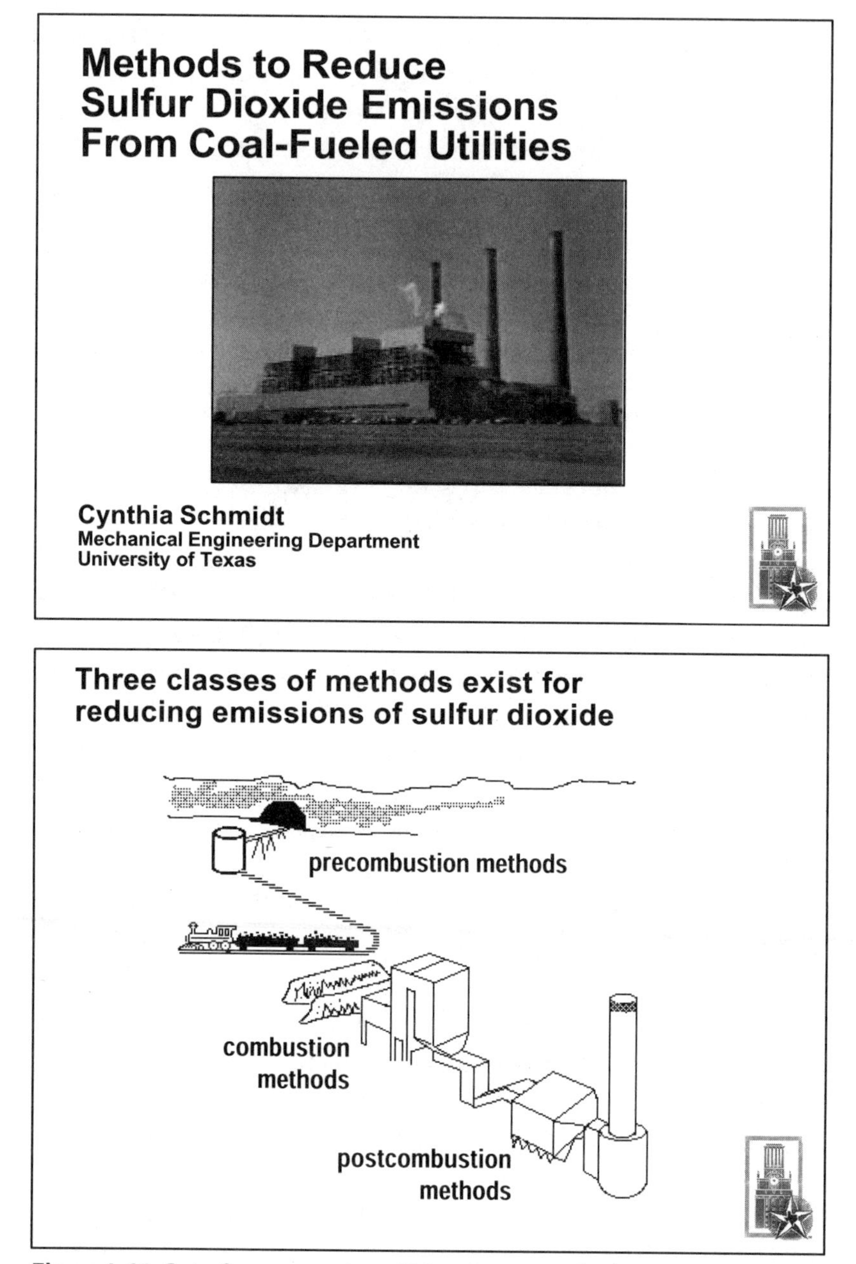

Figure 4–21. Set of presentation slides that reveals organization of the presentation: title slide (top), mapping slide (bottom), body slide for each of the three main parts of the presentation (top right, bottom right, and top of page 150), and conclusion slide (bottom of page 150).[9]

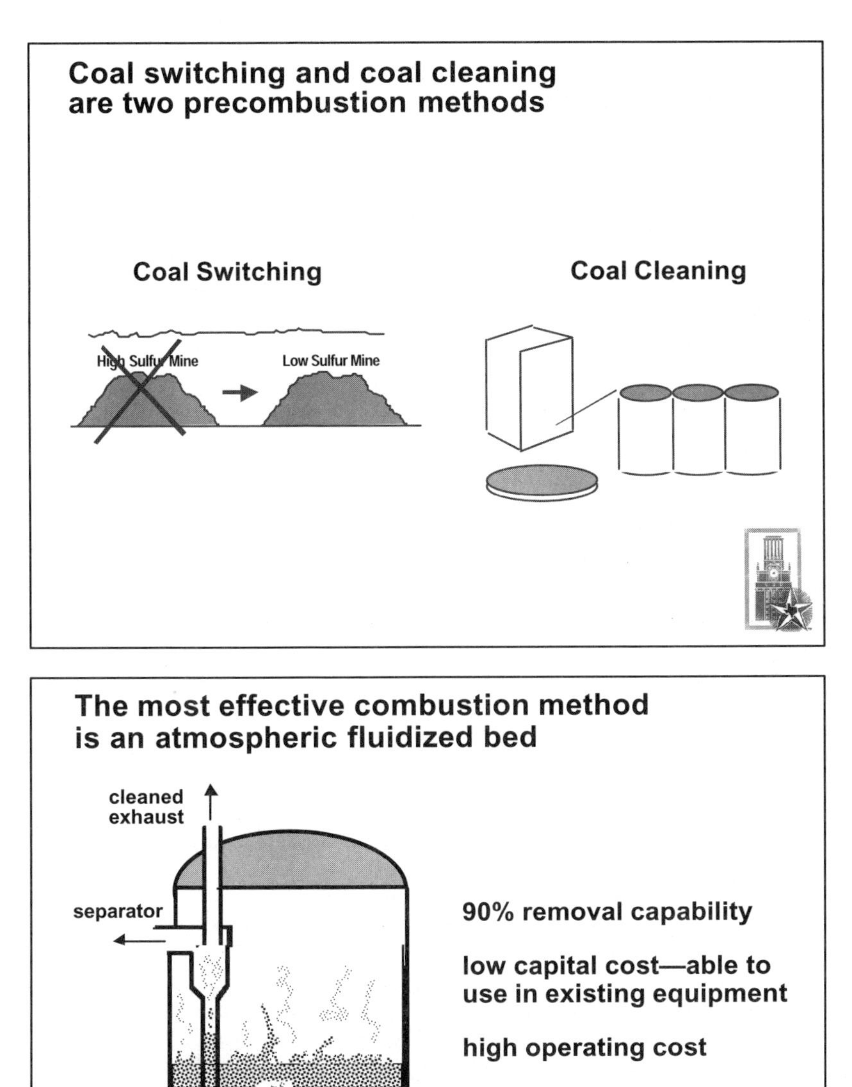

Figure 4–21 (Continued).

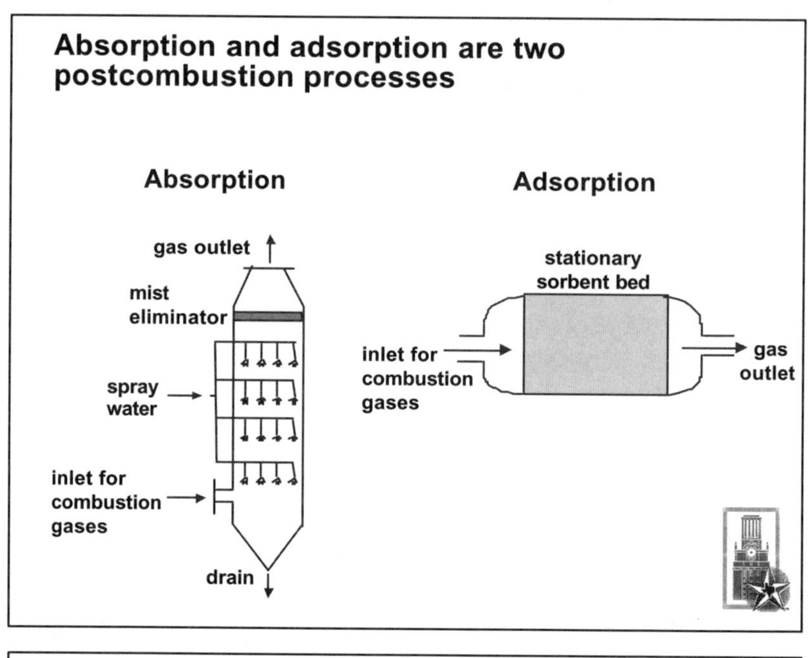

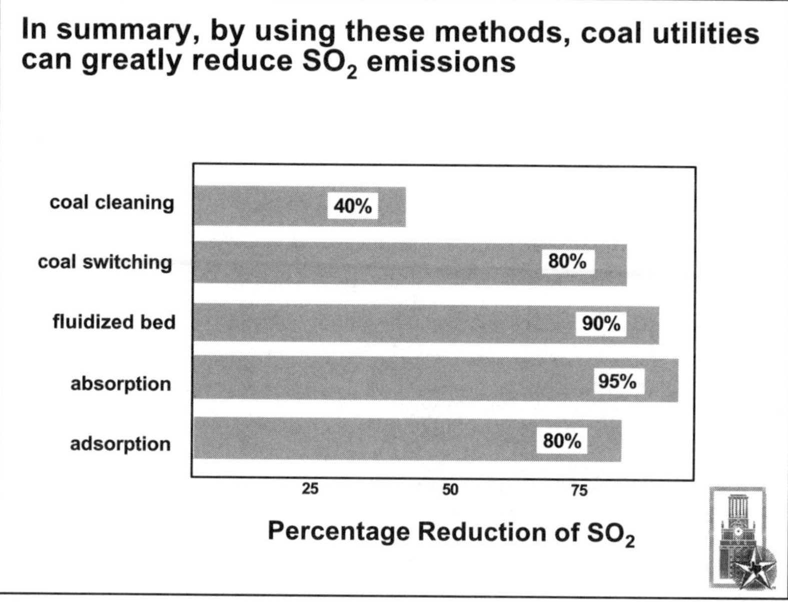

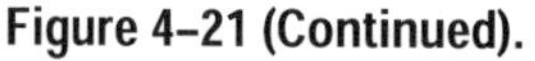
Figure 4–21 (Continued).

the mapping slide is projected for only a few seconds, which is not nearly enough time for the audience to memorize the list of topics.

Other key slides that reveal the organization are the first slides for the different parts of the middle. This presentation's middle was divided into three categories of methods. For that reason, the first slide for each category should signal the audience that a major transition is occurring. These three slides, also shown in Figure 4-21, do so both with repeated words (precombustion methods, combustion methods, and postcombustion methods) and with repeated images. For a long section, the first slide is often a mapping slide for that section, as in the case of the first slides for the section on precombustion methods and the section on postcombustion methods.

A final organization slide is a conclusion slide. The conclusion slide should help signal the audience that the end of the presentation is at hand. In addition, the conclusion slide should emphasize key results from the presentation. In the sample presentation of Figure 4-21, the conclusion slide emphasizes results from the most important criterion of the evaluation: the effectiveness of each method at reducing sulfur dioxide emissions. Granted, other criteria such as cost exist, but the results of those secondary criteria could be repeated in speech so that the results of this most important criterion receive the most emphasis.

On their conclusion slides, many presenters unfortunately resign themselves to using bulleted lists. As shown in Figure 4-22, the conclusions can be presented in a more memorable fashion. This presentation compared two computational methods for simulating the way a person can detect sound from a vibrating structure. The two methods (the exhaustive method and the singular value decomposition (SVD) method) were compared with respect to how many computations were required and how accurate the simulations were. The use of the balance scales on this conclusion slide made the comparisons more memorable than if the presenter had simply listed the two results. By the way, when this presenter, Aimee Lalime, was ready for questions in this presentation, she had the single word *Questions* appear at the bottom of this slide. That strategy allowed her to keep projecting the conclusion slide for the duration of the question period rather than to project a throwaway question slide as so many presenters unfortunately do.

For a short presentation, such as at a conference, you should limit yourself to one conclusion slide. Once the audience sees or hears "conclusion," they assume that they are at the end. As my wife contends,[11] a second conclusion slide tests the audience's patience, and the unthinkable third conclusion slide exhausts it.

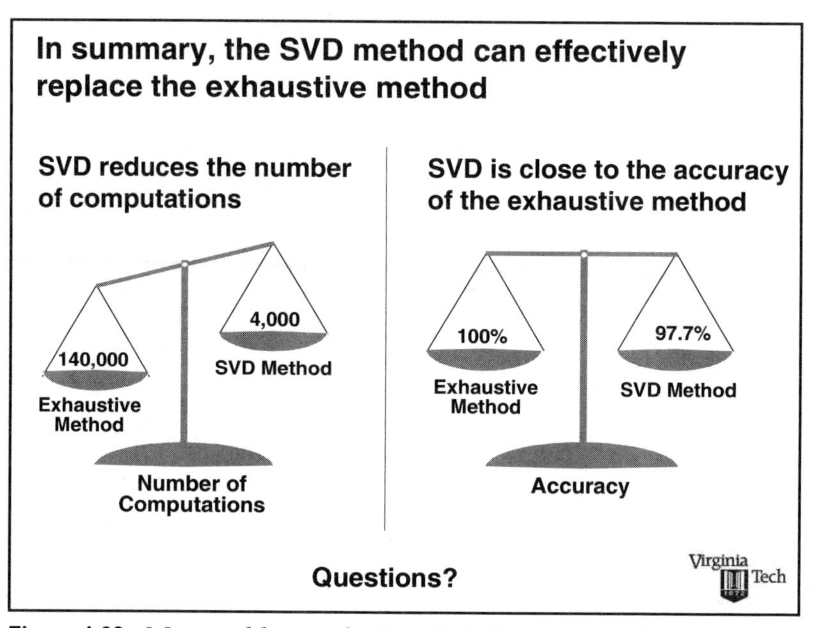

Figure 4-22. Memorable conclusion slide for presentation that compares two computational methods: the exhaustive method and the singular value decomposition (SVD) method.[10] The methods can be used to simulate the way we hear sound from a vibrating structure.

Critical Error 7
Not Accounting for Murphy's Law

Today I start my experimental lectures. They are also lectures of anxiety because with the setting up of every experiment comes the feeling: Will it go according to plan? For the best preparations are by no means an absolute safeguard against nature's perversity.[1]

—Heinrich Hertz

The legend goes that a new physics professor at a Midwest school wanted to impress the undergraduates with his teaching.[2] In his first semester, he requested to teach one of the large sections of freshman physics. Held in a huge lecture hall, each class period consisted of a lecture about a major topic and then a demonstration of a principle associated with that topic. For the class period that discussed motion in a plane, the new professor refused to use the six-foot air table that everyone else used. Instead, he had the technicians build him something twice as large. Likewise, for the class period on Newton's laws, he refused to use the tabletop spring balance that everyone else used; he had the technicians build him something much larger.

When the time came for him to demonstrate the motion of a pendulum, he put aside the tabletop pendulum that everyone else used and called for a hook to be mounted in the lecture hall's ceiling. Then he had tethered to the hook a medicine ball: one of those heavy leather balls that athletes in the 1950s tossed around for exercise. The medicine ball was fixed so that it could swing freely from one wall to the other. The day for the pendulum class came, and the new assistant professor began by saying that he was to perform an experiment that would place himself in physical danger. Although he said that he could be seriously injured in the experiment, he claimed that he had not practiced it because he was so confident that the laws of physics would keep him safe.

So, after preliminary discussion about a pendulum's trajectory, its period, and its minimum and maximum speeds, the new professor pulled the medicine ball to one wall of the room and climbed on top of a stepladder. With his back against the wall, the professor held the medicine ball to his chin and said that he was going to release the ball with zero velocity and that when it returned to this position, by the laws of physics, the velocity would again be zero. Well, this new professor's goal had been to engage the students, and as he stood on top of that ladder with the huge ball against his chin, he certainly had them engaged. When the professor let the ball go, it swung through its arc attaining a maximum velocity at the low point of the arc and rising almost to touch the far wall. Then the ball started its return trajectory. Apparently, because the new professor had not practiced this demonstration, he was unprepared for the sight of the huge ball making its way back toward him. According to students in the room, although his eyes grew large, he refused to bail out. Instead, he braced himself, but in so doing, he must have leaned forward ever so slightly. What the new assistant professor ended up demonstrating was not so much the motion of a pendulum, but the conservation of momentum because the students saw that a large mass (the medicine ball) with little velocity struck a small mass (the new assistant professor's head) and imparted to it a relatively fast velocity. The new assistant professor's head snapped back and hit the wall, and he fell unconscious to the floor.

For a moment, no one in the class moved. Then a few students rushed down to the new professor's aid. Later, after the smelling salts arrived and the professor returned to consciousness, the class slowly wandered out.

This professor's demonstration followed the Law of Murphy, which was named for Edward A. Murphy, Jr., an engineer who worked on rocket-sled experiments for the US Air Force in 1949.[3] Over the years, Murphy's Law has taken on many forms. With regard to presentations, the most fitting form is, "What can go wrong will go wrong, and at the worst possible time." Examples of Murphy's Law abound in presentations. One example occurred in a demonstration by Microsoft Corporation of its Office XP version of PowerPoint. The presentation occurred before technical professionals and government workers packed into the MCI Center in Washington, D.C. During the demonstration, the program crashed, locking up the computer.[4]

Yet another instance of Murphy's Law reigning during a presentation occurred with an architectural firm that had bid on the design of a new baseball stadium in Milwaukee. The firm, which was based in Milwaukee, had a design similar to the sky dome in Toronto in which the roof could retract on sunny days and close on rainy days. In the presentation of the proposed design, the firm set out to demonstrate the roof's movement on its three-dimensional model of the stadium. This demonstration was planned for the culmination of the presentation and was accompanied by "The Star Spangled Banner." With the music playing loudly and the audience focused on the beautiful model of the stadium, the presenter flipped the switch for the roof to open. Nothing happened. "The Star Spangled Banner" continued to play, and the audience continued to keep its focus on the beautiful model, but the roof did not retract. The presenter tried everything that he could, but the demonstration failed. As you might expect, in the stiff competition for the contract, this architectural firm did not win the bid.[5]

Not only does Murphy's Law wreak havoc during demonstrations, but it also causes mischief while presenters use equipment for projecting overhead slides. Consider, for example, the review meeting in St. Louis in which an engineer began his thirty-minute presentation by placing a transparency on the overhead projector. Unknowingly, the presenter had left the onionskin paper, which was not transparent, on the transparency.[6] Because the meeting was a review meeting and because everyone was competing for the same pot of funds, people in the audience were not inclined to help one another. For that reason, no one said anything to the presenter about the fact that nothing was projected onto the screen. The presenter did not help his own cause, because he stood beside the overhead, stared down at the sheet of paper, and never once looked behind him to see what was projected. Using a pointer, the presenter went line by line through a set of equations on the sheet. Nearing the bottom of the sheet, he said, "As you can see...." With that remark, chuckles sounded throughout the audience.

The presenter paused, looked up at the audience, chuckled himself, and then proceeded with the second sheet of his talk. Again, the presenter stared at the sheet and never looked behind at the screen. A few more times he said, "As you can see...," and each time the remark elicited more laughter. So it went for about fifteen minutes. Each time he said, "As you can see," the audience laughed, and each time he paused, looked up, laughed himself, and then proceeded. Finally, after a dozen sheets, he stopped and asked the audience, "Okay guys, what is so funny?" Someone yelled, "Take a look at the screen." He turned around, and saw that the screen was projecting a big black shadow.

Had he spent so much as thirty seconds before the presentation trying out his slides on the overhead projector, he would have realized his error. Although the step seems so obvious, the number of presenters who botch the execution of this simple piece of equipment is astounding.

Consider as another example the engineering professor who made a presentation at a Ohio review meeting of industrial sponsors. When the engineering professor placed the first overhead slide onto the projector, the slide was backwards and upside down. This audience, which was more helpful than the previously mentioned St. Louis audience, kindly informed the professor about the error. The professor then picked up the transparency and turned it right side up, but it was still backwards. The audience chuckled. On the third try, the professor placed the slide correctly. With the next transparency, the professor made the exact same mistake, and again it took him two iterations to correctly place the slide. This time, the audience did not chuckle. Although it is difficult to believe, the professor made that same mistake on all twenty of his presentation slides. In the middle of this presentation, one of the sponsors turned to another and said, "I swear, if that idiot makes that same mistake again with the slides, I will make sure that his contract is not renewed." Well, the professor continued making the same mistake, and the contract was not renewed.

Yet a third and final story (actually, I have many more) occurred at a briefing about a research funding opportunity. Using an overhead projector, but standing directly in its light, the funding agency's new representative went over each slide, point by point. Because the lists on her transparencies contained a rehashing of last year's specifications and because the audience was tired and somewhat jaded, no one told her that the entire projection was displayed in an undecipherable image on the front of her red dress. This representative continued in this manner for about fifteen minutes, when a latecomer to the briefing shouted out, "You are blocking the projector." The representative apologized and then moved about 6 inches to the side. She finished her presentation while now blocking about three-fourths of the screen, which the uninterested audience continued to ignore.[7]

The purpose of relating all of these accounts of presentations being pulled down by failed demonstrations or mishandled equipment is not to dissuade you from incorporating demonstrations or projected slides. Rather, the purpose is to make you sensitive to the hurdles of their incorporation. Your decision as to whether to incorporate demonstrations or slides should account for three factors: (1) the complexity of the demonstration or the difficulty in handling the projection equipment; (2) the gain for the presentation should the demonstration succeed or the slides project; and (3) the loss for the presentation should the demonstration fail or the slides not project. If the demonstration is worth trying or if the slides are worth projecting, you should go forward. In that case, though, you should find ways to minimize the occurrence of Murphy's Law.

Rehearsing

One of the best ways to minimize the occurrence of Murphy's Law in your presentation is to rehearse. For each of his spectacular high-voltage demonstrations, Nikola Tesla reportedly rehearsed at least twenty times.[8] By rehearsing, you learn the pitfalls that could startle you in the actual performance. For instance, had each of the three presenters who mishandled the overhead projector simply practiced for a couple of minutes, each presenter would have avoided the mistakes that pulled down the presentation. Likewise, if the Midwest professor had practiced with the medicine ball mounted on the pendulum, he would have realized the problems inherent in that demonstration (namely, his own fear of the ball) and reworked the experiment.

Rehearsing is certainly no guarantee of success. The architectural firm that was competing for the contract for the baseball stadium had practiced opening and closing the roof several times before the big proposal presentation. Moreover, Microsoft had undoubtedly practiced the demonstration of its XP software several times before its presentation in the MCI Center. Nonetheless, while rehearsing does not guarantee success, it greatly increases the odds.

For instance, during one rehearsal with my computer projection system, which has a remote control to change the visuals, I learned two important lessons. First, if the laptop computer is moved during the presentation, the antenna for the remote can become disengaged. Second, I learned that simply plugging the antenna back in does not cure the problem—the slides continue to switch, but do so at a painfully slow rate. To get the system to work effectively again, I have to stop the slide show of the presentation program and then remove a hidden computer window that warns about the antenna having been disconnected. Learning these two lessons has saved me much grief and embarrassment in my presentations. I am careful to make sure that the connection between the laptop and antenna is secure, and on the one odd occasion when someone moved my computer and inadvertently disconnected the antenna, I efficiently took the steps necessary to get the system back on track.

Arriving Early

Another important step to mitigate the effects of Murphy's Law is to arrive early to the presentation. When I taught at the University of Wisconsin, I had to give a large lecture each semester in an auditorium to about 150 upperclassmen. This room had a projection system that operated from a computer within the room, and each semester I would carry over two computer disks (I brought two in case one failed). For the first four semesters, everything worked well. Although I revised the slides each semester (I continually revise my presentations), all four presentations went smoothly.

For the presentation in the fifth semester, though, I was thrown for a loop. When I loaded my presentation onto the auditorium's computer and opened the file, I was shocked to see that each letter of my presentation had been replaced by a little box. In other words, the typeface that I had used for this presentation (**Arial Narrow**) had been removed from the machine. The audience, which was already filtering in, did not know my dark secret, because this wonderful room was equipped such that you did not have to project what was on the computer into view until you wanted to. Because I had arrived ten minutes early, I had the opportunity to highlight all those slides and replace my typeface with one on the machine, **Arial**. Because **Arial** is significantly wider than **Arial Narrow**, I had to finagle some of the text boxes. Although that took a little time, I was ready to go when the bell rang and the audience expected me to deliver.

What saved me from embarrassment in that presentation was my early arrival to the lecture hall (granted, bringing my own laptop or bringing my own fonts on the disk would have circumvented the problem). By the way, in that same lecture hall, I have watched presenters irritate audiences because these presenters arrived just as the bell rang only to find that something unexpected had happened. In some cases, their typefaces had been removed from the computer as mine had been. In other cases, their computer disks did not work. In still other cases, their laptop computers did not have the right connections with the projector. In all those cases, the speakers taxed the patience of a hundred or so members of the audience. Few scenes in scientific presentations are as painful as the one when the moment for the presentation arrives, the large crowd quiets down and focuses its attention on the speaker, and the speaker is frantically working on his or her computer, unprepared to begin.

Accounting for the Worst

Usually, a few days before a presentation, when I have my structure and projected slides set, I imagine what I would do if the worst were to occur. Often, I imagine this nightmare while I am taking my noontime run or walking my dogs. In imagining the worst, I am not psyching myself out by dwelling on failure; rather, I am trying to devise a plan should the equipment fail. Such thinking is good preparation.

For instance, in one presentation before seventy-five people at one of the national laboratories, I had requested a computer projection system. This presentation occurred when the technology for computer projectors was new. I was skeptical about the equipment working because I had never used such a piece of equipment at this laboratory. For that reason, I had designed the presentation such that I could give it from just my handouts. Sure enough, the unexpected happened: The computer technician went on vacation the day before I arrived to speak, and the backup person did not receive word about my request until five minutes before the presentation. For the seventy-five people crowded in the room, I began the presentation on time using the handouts. Fifteen minutes later, the computer was up and running, but those important minutes, as well as the patience of the audience, had not been lost.

Imagining potential problems is a good exercise, but imagining the worst is even better. I learned this lesson while conducting a six-hour workshop at a different national laboratory. The first problem that occurred was that some handouts I requested to be shipped to the presentation site in California came instead to my home in Virginia. This hurdle was not such a big problem because I could simply pack those handouts in a suitcase. However, because my suitcase was now too large for me to carry onto the plane, I had to check it through. The second problem was that the airline misplaced that bag, and it did not arrive in San Jose, California, with my flight. Because the presentation was the next day, I had to begin the presentation without the handouts. I also had to make the presentation in the clothes that I had worn on the plane; I had arrived late at night when all the clothing stores were closed. Making the presentation in the same clothes that I wore on the plane was not such a problem, because for the plane trip I had worn professional attire. Granted, the clothes did not feel fresh, but that was more my perception than the audience's.

The second problem that I encountered for this presentation was that the on-site computer projection system had a burned-out bulb. I had never used a projector such as this one, which was in a big black box with a strange cable arrangement. Fortunately, as is my custom, I had arrived thirty minutes before the presentation and had time to change the bulb. Unfortunately, even with a changed bulb, the projector did not project my slides, because it was not receiving a signal from my computer. Apparently, this kind of projector required special software to work with my laptop computer. As luck would have it, the site had a second projector down the hall, this one of a different type. Unfortunately, someone had walked off with the cable that was to connect to my computer. For emergency situations such as this one, I had brought a few transparencies so that I could begin my presentation. Because the time for the presentation had begun and the room was filled, I began my presentation with these transparencies as my host from the laboratory frantically searched for a cable. I spoke slowly because I had only one hour's worth of transparencies for a six-hour workshop.

After fifty minutes, when I was down to only a couple of transparencies, my bad luck changed. My host found a third projector, this one with a cable and without the need for special software on my computer. So I was able to switch back to my computer. Moreover, during the lunch break, the airline reported that they had located my bag. With ten minutes to go in the workshop, the airline delivered the bag—just enough time for me to distribute the handouts to the participants before they left.

My assessment at the end of this workshop was that I had been lucky.

Disasters usually do not occur for just one reason but for a series of reasons. Consider a case much more serious than a failed presentation: the sinking of the *Titanic* and the loss of more than 1500 people. In the *Titanic*'s case, the reasons for the disaster were numerous: The captain had apparently wanted to set a speed record for the voyage; the sea was unusually calm, so that the lookouts could not see waves lapping against the iceberg; the lookouts had misplaced their binoculars and had to rely on their unaided eyes; the pilot did not hit the iceberg head on (which many believe would have allowed the *Titanic* to stay afloat for several hours), but hit it with a glancing blow that caused much more damage to the hull; the crew had not practiced filling or lowering the lifeboats; earlier in the evening, the wireless operators of the *Titanic* had chastised the wireless operator of the *California*, the nearest ship, for sending them a warning about the ice (the *Titanic*'s operators were busy sending messages to New

York); the *California*'s wireless operator, rebuffed by the *Titanic*'s operators, went to bed early, less than an hour before the striking of the iceberg.[9] Such a string of events could cause even the best prepared presentation to fail.

Although you might rehearse, arrive early, and anticipate the worst, you will probably encounter in your career at least one set of circumstances in which Murphy's Law will reign. In such a situation, you should keep your cool and, as Michael Faraday did, keep control of the situation. Michael Faraday performed many experiments in his lectures and therefore took many risks. His skill at experimenting impressed even the best scientists of his day, including Joseph Henry. Still, Faraday's experiments were not immune to Murphy's Law. As Faraday's biographer Geoffrey Canter commented, Faraday was in "apparent total command of himself and therefore of the proceedings. This is not to say that experiments did not sometimes fail to function as expected, but on such occasions he could turn the apparent failure to advantage and not lose control of the situation."[10]

Chapter 5

Delivery: You, the Room, and the Audience

[Feynman] absolutely riveted the attention of everyone in the room for the entire time he was there. His need to do that helps explain some of the racy stories he liked to tell about himself, but it also lies close to the core of what made him a great teacher. For Feynman, the lecture hall was a theater, and the lecturer a performer, responsible for providing drama and fireworks as well as facts and figures. This was true regardless of his audience, whether he was talking to undergraduates or graduate students, to his colleagues or the general public.[1]

—David L. Goodstein

Delivery is your interaction with the audience and with the room. Voice, gestures, eye contact, stance, movement—all of these contribute to delivery. How you deliver your presentation affects how intently the audience listens to you and whether your audience even trusts you. According to Michael Faraday, "[Lectures] depend entirely for their value on the manner in which they are given. It is not the matter, not the subject, so much as the man."[2] What Faraday meant here was not that the quality of the content was unimportant, but that no matter what the subject is, the audience will be engaged only if the speaker delivers that subject in an engaging way. For instance, over the years, one of the most popular courses at Cornell has been beekeeping. Is that because so many students attend Cornell because they want to become apiarists? No. The reason for the course's popularity has been that the faculty are so passionate about beekeeping and know the subject so well that students naturally have become interested.

Different Styles of Delivery

So, for what kind of delivery should you strive ? The answer to this question is not simple. Certainly, Richard Feynman provided an excellent model. He enthralled audiences—not an easy task when one's topic is as technical and abstract as Feynman's studies on quantum electrodynamics were. Linus Pauling also affected audiences in this way. However, not all of us are suited to deliver with the charisma of these two.

Several presenters have influenced the way that I give presentations. Two of these are Kamalaksha Das Gupta, a former physics professor at Texas Tech University, and Patricia Smith, a director at Sandia National Laboratories. Das Gupta and Smith reveal a stark contrast in delivery styles. Smith is the consummate professional: well prepared, poised, and appropriately dressed. One of Smith's strengths is how well she handles questions, even the caustic ones. In handling difficult questions, Smith maintains a calm, but resolute, voice. By keeping her cool, she shows that she, not the caustic questioner, controls the presentation. Also, by methodically considering several different perspectives to the concern raised, Smith refuses to be boxed into the *either–or* traps that such questioners often lay.

Das Gupta is quite different in his delivery from Smith. Das Gupta, who studied x-ray physics under the great Bose, probably breaks every prescribed rule of dress, eye contact, and stance. He wears sandals, no socks, and a green plaid jacket caked in chalk dust. As he talks through difficult points, he often closes his eyes, leans against the blackboard, and presses a fist against his forehead. Despite giving the appearance of being self-absorbed, Das Gupta actually has a keen sensitivity to his audience and can make sense of the most poorly phrased question. What distinguishes Das Gupta, though, as a presenter is his sincere passion for his subject and his deep knowledge about that subject.

The delivery style of Das Gupta falls more in the category of Feynman's. Das Gupta's personality is more suited, though, for that type of delivery. Conversely, the delivery style of Smith is more low-key, which is more like the delivery style of Lise Meitner. In developing your own style of delivery, you should reflect on what kind of delivery you feel comfortable giving. For instance, do you prefer to move around before the audience or are you more at ease standing behind a podium?

One's personality is not the only thing that shapes one's style of delivery. Also affecting delivery style is the audience. For instance, standing before an audience with whom I am comfortable and who boost my confidence with smiles and nodding heads, I walk around, tell stories, take chances with humor, and vary the loudness and pitch of my voice. However, standing before an audience with whom I am not familiar or who are decidedly antagonistic, I am much more serious and businesslike. I do not vary my movements as much. Nor do I vary as much the loudness or pitch of my voice. As mentioned in Chapter 1, Rosalind Franklin faced an antagonistic audience in a 1951 presentation of her x-ray crystallographic work on DNA. Given the hostility that she faced, James Watson's criticism of her lack of warmth was unfair.[3]

Not only does the sentiment of the audience affect your delivery, but so does the size of the audience. Although I prefer to move about the room, tell stories, and engage the audience during my presentations, I have found that with larger audiences this approach is risky, particularly at the beginning of the presentation. The reason is that the larger the audience, the more likely that I will miss individual signals of distress. If some people are confused by something and I do not pick up that confusion in those people, then I do not realize that I should reword my point. Because I am presenting in a flamboyant style, the audience is bothered more by the confusion than if I were presenting in a humble style. The confusion festers like a sore and comes out during the question period in an antagonistic way.

Yet a third effect upon one's delivery style is the occasion. While many occasions allow for a delivery that is animated, some occasions call for a more somber delivery. An aspect of occasion that significantly affects one's delivery is the room. A presenter's delivery in a cozy conference room differs significantly from the delivery in a lecture hall with vaulted ceilings and tiered seats. In the large lecture hall, a formal barrier exists between the speaker and the audience. Although the speaker can engage the audience by moving up to the people seated in the first rows or along the aisles, the barrier poses a formidable challenge. In addition to the size of the room, an equally important consideration is how filled the room is. Engaging an audience in a half-filled room is much more challenging than engaging an audience in a filled room. In a filled room, if you happen to say something witty and the audience responds with laughter, that laughter fills the room. However, in a half-filled room, the laughter quickly dissipates. Yet another consideration for the room is its layout. If you desire discussion among the audience, a U-shaped seating arrangement works much better than a room with seats arranged in rows, because in a U-shaped arrangement, the audience can see one another.

Opportunity to Improve Delivery

Delivery is an aspect of presentations in which you can make marked improvement with conscious effort. For instance, the initial lectures of Heinrich Hertz were marked by his avoiding eye contact with the audience. In those initial lectures, he looked either at the blackboard or down at his paper. Moreover, not only did he read his presentations, but he read them quickly. In fact, at a job-interview lecture at Kiel, every member of the evaluation panel commented that Hertz had spoken too quickly, a sign that he must have raced through his talk, because in most audiences, at least one person is too polite to criticize. At that particular lecture, Hertz did not have a clock and was concerned about keeping his talk within the time limit of 45 minutes. Unfortunately, the more concerned he became about the time limit, the faster he spoke.[4]

However, Hertz's delivery improved with time. Interestingly, his self-critiques of his later presentations differed significantly from his earlier self-critiques. When writing about his first few lectures, he showed the false assurance that one has when he or she knows that things have not gone well but does not know how to correct the matter.[5] However, when writing about a later presentation (his inaugural address at the University of Karlsruhe), he criticized himself harshly, even when his audience had given him only praise. In a letter to his parents, he wrote the following about that lecture:

> Yesterday I shook off at least one threat, the inaugural lecture. The professors nearly all attended, as did *Ministerialrat* Arnsperger and *Staatsrat* Nokk (the minister of education). My

> speech did not satisfy me at all. In my opinion it failed terribly (you will get a chance to see for yourselves). Likewise the manner in which it was presented left more to be desired than strictly necessary. The fact that I nevertheless heard only kind reactions only shows the modest expectations of the audience.[6]

No doubt Hertz had improved from his earlier presentations, but no doubt he had established much higher goals for himself.

Hertz's pinnacle for presentations occurred in a lecture given to more than three hundred spectators at the Polytechnic in Bonn. At the end, "there was so much applauding and cheering" that his wife became embarrassed. Afterwards, one man wrote to say that he was quite "shaken" by the lecture; another claimed that although the lecture "cost him a sleepless night," he did not regret it.[7]

So how does one improve one's delivery? Hertz's improvement from the nervous student in Berlin to the composed lecturer in Bonn occurred over nine years (1880–1889). His method was one of continual self-criticism and revision. For instance, in 1883, he reduced the notes for one of his lectures by almost one-half and slowed his tempo. Afterwards, he worried that he had spoken too slowly, but his audience of professors and students assured him that the pace had been "quite right."[8] What is important to note here is that Hertz made the effort to query his audience—a sign that he desired to improve.

For most of us, nine years is a long time to improve our delivery, even if we were to achieve the dramatic changes that Hertz experienced. How then could we improve our delivery in a shorter time?

One of the best ways is to have colleagues critique your presentation. These critiques should not only mention those aspects that are weak, but also discuss what aspects of delivery were strong (for a critique sheet, see Appendix A). Also effective is to videotape yourself making a presentation and then to review the videotape with a critical eye. Although this exercise can be frightening, it is enlightening to see your movements and expressions and to hear your voice.

In improving your delivery, one thing to be careful of is that you do not try to orchestrate every individual part of the delivery: posture, stance, hand movements, body movements, facial expressions, eye contact, loudness of voice, variation in voice, and avoidance of filler phrases. If you consciously worry about every one of these aspects, you might neglect bigger things such as sincere enthusiasm that are needed for success. In other words, in trying to improve all the aspects of your delivery, you might become so stiff and self-conscious that your presentation does not engage the audience.

In my experience, the most captivating speakers in science and engineering have been the ones who loved their subjects and knew them well. If you do not convey your interest for the subject, how can you expect your audience to become interested? Conveying enthusiasm does not mean that you sing and shout. If you put on pretences, the enthusiasm will not be sincere, and the audience will see through the act. Moreover, scientific audiences are suspicious of deliveries that contain too much dazzle, particularly at the beginning; these audiences often assume that the presentation is more style than substance. In truth, a presenter can show a genuine enthusiasm for the subject in many ways: a sincere voice, sustained eye contact, animated facial expressions, and natural gestures that contribute to the audience's understanding of the subject.

Another way to improve your delivery in a relatively short time is to study the delivery of others with a critical eye, as Michael Faraday did. Nobel winner Rosalyn Yalow also adopted this technique before her first teaching assignment. According to Yalow, "Like nearly all first-year teaching assistants, I had never taught before—but unlike the others, I also undertook to observe in the classroom of a young instructor with an excellent reputation so that I could learn how it should be done."[9]

In using this technique, you should try to imagine someone who is an excellent speaker making your presentation. Imagine the rhythm of his or her voice. Imagine his or her movements on stage. Ideally, this imagination should occur soon before you go on stage.

This strategy for learning to deliver a presentation follows the advice of Tim Galloway[10] for learning to play tennis. Rather than having his students worry so much at the beginning about how to grip the racket, position the feet, bend the knees, and address the ball, Galloway has his students simply watch a videotape of a great tennis player making a series of forehand shots, backhand shots, volleys, or serves and then has his students go out onto the court and do likewise. Even though I learned to play tennis years ago, I continue to use this strategy. For instance, just before hitting a topspin forehand, I imagine Steffi Graf hitting a topspin forehand. On television, I have watched Graf hit hundreds of such forehands, and so in my mind I replay the images of her moving to the ball, positioning herself, and stroking the ball. And then I try to do likewise. With speaking, I do a little of the same. I imagine one of my model speakers—Patricia Smith or Kamalaksha Das Gupta—and try to emulate that person's style. Having these individuals as models does not diminish my individuality as a speaker. Rather, it helps me bring out those traits in my own delivery that I value so highly in theirs.

Critical Error 8
Not Preparing Enough

A young [person] doesn't realize how much time it takes to prepare good lectures, for the first time, especially."[1]

—Richard Feynman

After his work at Los Alamos during the war, Richard Feynman began teaching at Cornell. For a long time, he became depressed at how little research he was doing. He claimed to feel "tired"[2] and was unsure what was making him feel that way. Later, he realized how much time and energy preparing his class presentations required. Years later, because of the time that preparing a new lecture demands, Feynman hesitated to do the set of freshman physics lectures that ended up bringing him so much acclaim. As he predicted, the freshman lectures did consume his time, causing him to put aside his research. Although he spoke only twice a week, he "worked from eight to sixteen hours per day on these lectures, thinking through his own outline and planning how each lecture fit with the other parts."[3]

Preparing a strong presentation does takes time. Time is needed to understand the content well enough to organize it in a fashion that is readily comprehended by the audience. Time is also needed to gather the important images, to graph the important results, and to incorporate those images and graphs into a set of well-designed slides. Moreover, time is needed to rehearse the material so that the speaker can find the right words to explain the difficult concepts and to smooth the transitions between points.

Preparing Visual Aids

Early in the preparation process for the presentation you should consider the visual aids that you want to use. Spending time early on visual aids is important, because preparing these aids consumes much time. For instance, to craft a set of presentation slides, you need time for gathering the images and formatting the graphs. Then there is the time to design each slide. Then you need time to organize and format those slides so that as a set they reflect the talk's organization. In doing so, you will have a title slide, perhaps a background slide, a memorable mapping slide, slides for each of the talk's main divisions, and a conclusion slide. Yet more time is needed for you to rehearse with those slides so that you can make the appropriate transitions. Finally, if you are to give a version of those slides as handouts, time is needed for creating that handout version, printing it, and then photocopying it.

When the presentation is collaborative, even more time is needed to prepare a set of presentation slides, because each speaker should have the chance to comment on a draft of those slides. Ideally, in a collaborative presentation, one person should have the task of creating the slides. That scheme makes it easier for the group to obtain both a consistent slide format and a set of slides that reveals the presentation's organization.

To prepare a film, time is needed to prepare the film, either through photography or through computer simulation. Then you have either to incorporate the film into a computer projection or to prepare a videotape of the film that can be shown. As with the slides, you should allot time for practicing. To prepare demonstrations, a similar schedule is needed: design, construction, and practice.

For visual aids, one additional step is needed: development of a backup plan in case Murphy's Law reigns.

Most likely, sometime in your career, you will have to rely on your backup plan. More than once, I have traveled to a site to make a computer projection, only to find that something was not as expected. One time a bulb burned out in the middle of a presentation, and there was not a spare handy. Another time, I had a projector that worked, but the cable to my laptop did not have the right connector. Yet another time, I had a projector that worked, I had a cable with the right connectors, but my laptop needed to have special software to run the projector. Still another time, I had a computer, cable, and projector, but the computer accepted only a CD and all I had was a zip disk. Fortunately, on each occasion, I resorted to my backup plan, which for each of those cases happened to be overhead transparencies.

Preparing Yourself to Speak

One day my wife lamented about how one of her graduate students had spent far too much time—the better part of six months—preparing a proposal of his doctoral work for his committee. "If he had put that time into his research," she complained, "he would be a good portion of the way finished with his project."[4] The communication requirements for this proposal were not so high: a five-page document and a twenty-minute presentation. This student, though, had written more than sixty pages and had prepared more than thirty-five presentation slides. This high number counters the rule of thumb back in Table 4-2 of dedicating at least one minute for each slide (and preferably at least two minutes for a slide with a key graph or complex image). The day before the presentation, my wife tried to persuade this student to stop working on the slides and to spend time rehearsing the presentation. Unfortunately, the student continued tinkering with the slides up to the hour of the presentation.

By the student's own admission, the presentation was a failure. He had problems, as he said, "finding a rhythm." Things never clicked for him, and he struggled to find transitions between his different points.

As mentioned in the previous section, a speaker needs time to practice, even if he or she has the best set of presentation slides. Practice helps the speaker with transitions from one point to the next. Practice also helps the speaker work through the explanations of difficult concepts so that all the words are, in fact, inside the speaker and ready to come out. Most important, perhaps, practice reassures the speaker that he or she can, in fact, make the presentation. Perhaps the greatest source of nervousness for speakers is the fear that they will stand before an audience and not know what to say. By having walked through the presentation, even if in a mumble, the speaker knows that the words are there.

To prepare themselves to speak, many presenters require some time alone before the presentation. For example, when Heinrich Hertz began teaching, he claimed that he could think of nothing else but each lecture for at least one hour before he gave it.[5] According to one of her daughters, Marie Curie required the entire afternoon to prepare herself for her five o'clock lecture to her graduate students.[6] As mentioned earlier, in preparing his freshman lecture series on physics, Richard Feynman spent eight to sixteen hours a day preparing for the series. Feynman spent this many hours each day of the week, not just on the two days that he spoke.[7]

Preparing a Speech in Another Language

Whenever I begin teaching a short course in Barcelona, I always try to say a few words in Spanish. The night before the course, the time that I prepare for those five minutes of Spanish is close to the time I spend preparing for the remaining three hours of the short course. Speaking in a foreign language significantly increases the challenge of the presentation. Granted, much depends on how well one knows the language, but anyone who attempts to make a scientific presentation in a language different from his or her own deserves much respect.

In making a presentation in a different language, not only are your speaking skills important, but so are your listening skills. Listening is important for understanding questions, which can arise from several different people, each with a significantly different accent. For that reason, just memorizing and practicing a speech in the other language is not enough, as Niels Bohr found out in his meeting with Churchill (discussed in Chapter 2). You have to be able to understand the questions and to respond on the spot.

An undesirable situation often arises when someone tries to learn a foreign language. The person makes mistakes (as is natural), then feels embarrassed, and then avoids speaking. Becoming better at that language then becomes impossible, because to learn a foreign language you have to speak that language. Although you will make mistakes in speaking in a different language, there is no reason to be embarrassed by those mistakes. Although the physicist Chien-Shiung Wu never felt at ease with English, she did not back down from speaking it. After earning her Ph.D., she went on a lecture tour across the United States. In her presentations, Wu often confused the pronouns *he* and *she,* and left out articles from her sentences. Because of her struggles, she wrote out her entire presentations and practiced them repeatedly beforehand.[8] Still, Wu did not shy away from speaking, and her tenacity at continuing to speak before crowds served her well in her career.

Critical Error 9
Not Paying Attention

Since we couldn't understand what [Oppenheimer] was saying we watched the cigarette. We were always expecting him to write on the board with it and smoke the chalk, but I don't think he ever did.[1]

—James Brady

In describing how he felt giving a paper to a geological society, Charles Darwin said, "I could somehow see nothing all around me but the paper, and I felt as if my body was gone, and only my head [was] left."[2] The sense of being disconnected that Charles Darwin experienced reflects the way that many presenters carry themselves during a presentation, as if they have no idea about the elements around them: the room, themselves, the audience, or the time.

Paying Attention to the Room

For years now, I have had dogs—large, outdoor dogs. As a rule, these dogs are unruly. They do sit, stay, and come, but only after hesitation. Moreover, they are restless creatures who in a few minutes of sitting in a veterinary waiting room can exhaust me with their squirming and pacing. One thing I have noticed, though, is that within seconds of my vet entering the examination room, they become still and attentive, almost subservient. So one day I asked my vet how she was able to exert that effect on these animals. The vet, whom I had known for several years, confided that much of it arose from her demeanor when entering the room. When entering, she did not make eye contact with the dog. Rather, she began setting up and rearranging things in the room. The dog was sizing her up at this point, and by taking control of the room, she let the dog know that this room was her room. Then, when she finally turned to the dog, it was with purpose. Dogs do not have much patience for being probed and pricked. So, when she attended to the dog, she did so with efficiency.

Although the audiences for scientific presentations are much more sophisticated than most dogs, we can learn much from this vet about how to approach a new audience. When an audience attends a scientific presentation, they want the time to be worthwhile. However, they have had so many empty experiences at scientific presentations that they fear the worst. For that reason, when you make a presentation to a new audience, show them early on that you mean business and that you will deliver. Granted, you should not be as cold to the audience as my vet first appears to my dog, but you should exhibit control of the situation. The lights, the arrangement of your speaking space, and even the arrangement of the seating for the audience—all of these are part of your domain.

So often I see presenters remain passive about these elements, much to the detriment of their presentations. For instance, because many speakers do not rearrange the front of the room, they often find themselves in awkward positions—on the wrong side of the overhead projector or boxed in by the furniture. Also, because many speakers do not check out the different possibilities for the lights, they end up projecting slides that are washed out or in rooms that are too dark for eye contact to be made. The advice here is simple: Take charge. After all, you are the one who will be credited or blamed for the presentation. So, if you prefer to walk around during your presentation, adjust the speaking area so that you can do so. If you prefer to stand on the left side of the overhead projector, then move the podium so that you can do so.

Granted, when you are at a conference and are speaking in a session with other presenters, you do not have as much freedom to rearrange the room as when you are doing a stand-alone presentation. Still, you should arrive early to your session, become familiar with the setup, and decide how best to work with the arrangement. Do not show up one minute before the session, as the opening speaker did for a session at a recent national conference. This speaker not only arrived late but demanded a different kind of projector from what the other four speakers were using. Hastily, he replaced the existing computer projector with an overhead projector. In doing so, he inadvertently closed the second speaker's laptop computer and caused it to go into a deep sleep. The result was that the beginning of the second presentation was delayed, and the second speaker had to cut short her talk.

As you are making the presentation, you still have the responsibility to exercise control of the room. For instance, if distracting noises are coming from an open door, take control and shut the door. If someone in the audience stands up to leave early, mitigate the disruption by looking to a different person on the opposite side of the room. If an outside disturbance occurs that is so loud that no one can hear you, stop speaking until the loud noise ceases. Over the years, I have witnessed speakers being drowned out by the rattling of heating pipes, the hammering from a laboratory, the emptying of garbage dumpsters, and on one occasion the roar of a passing train. Although you cannot control the train schedule or the sources for many of these noises, you can control your reaction to them.

Paying Attention to Yourself

Besides paying attention to your surroundings, you should pay attention to yourself: what you wear, how your voice projects, and how you move.

Attire. As a speaker, you can significantly influence the formality of an occasion by what you wear. Granted, scientists and engineers do not have the reputation for being well dressed. For instance, the first time that Einstein taught a university class, he arrived in "somewhat shabby attire, wearing pants that were too short."[3] Emmy Noether, the great mathematician, was also noted for having a disheveled appearance.[4] Likewise, the microbiologist James Watson once wore clothes purchased at an army PX to give a presentation at a conference in France; his clothes had been stolen on a train in Italy.[5]

On the other hand, Albert Michelson dressed formally for his class lectures in a "black square-cut morning coat, stiff high collar, and knife-edged, pinstripe trousers."[6] Even more impressive, Nikola Tesla wore a white tie and tails to make his presentations.[7] A professional appearance can give an audience a good first impression. That strategy was used by Nobel winner Rita Levi-Montalcini. To promote her work on nerve growth factors, Levi-Montalcini adopted an elegant and chic appearance for presentations.[8] Dressing in the flair of Italy, her native country, Levi-Montalcini showed up to presentations in a black sleeveless dress, of her own design, with a matching jacket, pearls, and four-inch heels.

Voice. Besides paying attention to dress, you should also think about your voice. Voice is a distinctive feature of a presenter. Ernest Rutherford, for instance, had a booming voice that was recognizable from the next room. Marie Curie had a soft but steady voice. Nikola Tesla had a "high-pitched, almost falsetto voice."[9] Einstein had an equally distinctive voice with a German accent. Although you cannot do so much with the pitch or accent of your voice, you can control the inflection and loudness. If your voice has no change in loudness or speed, you will quickly tire an audience. Heinrich Hertz, for instance, disliked meeting with Hermann Helmholtz, because Helmholtz spoke so slowly and deliberately that Hertz found it "impossible" for him to listen attentively.[10] James Watson also complained about the presentations at one international biochemical conference because there was "so much droning" that he found it difficult to "stay alert for the new facts."[11]

Changing the speed and loudness not only prevents the speaker from hypnotizing the audience, but it helps the speaker emphasize key details. The best speakers, Feynman and Pauling, changed their loudness and speed dramatically during a presentation. Such changes, though, should occur naturally; otherwise, the audience senses that the speaker is acting. In other words, the speaker should have the same voice inflections in loudness and speed that the speaker naturally has in conversation.

Movements. Equally important to paying attention to your voice is paying attention to your movements. These include your stance and the movements of your hands and feet. With your stance, you want to find a stance that conveys confidence to the audience and that makes you comfortable. Having your hands relaxed at your side conveys confidence, although many speakers find that stance unnatural. If there is a podium, you might try placing your hands lightly on the podium. Clenching the podium, though, conveys a defensive posture. On more than one occasion, I have seen a speaker clench the podium so tightly that veins bulged from the neck.

Besides wanting to exude confidence, speakers often want to convey that they are relaxed. A hand in a pocket conveys this demeanor, but the hand should not move. Such a movement distracts. Also, in regard to the pockets, remove your keys or change before the presentation. You might absentmindedly rattle them and distract the audience.

Some presentation books spell out a number of positions to avoid: both hands in the pockets, hands folded across the chest, a fig-leaf position (hands locked in front), reverse fig-leaf (hands locked in back), leaning against the podium, and so forth. In general, that advice is fine and well intended, but it should not inhibit your energy. In the middle of his presentations, Richard Feynman moved into a number of these positions—both hands in his pockets, for instance—but he adopted these positions only after he had engaged the audience, and these positions he held only briefly.

In addition to the way you stand, an important consideration is the way you move. The best presenters move during their presentations, but they move with purpose, and those movements contribute to the presentations. For instance, walking toward the audience can be a powerful movement that helps emphasize a point. Using your hands to illustrate points, as demonstrated by Karen Thole in Figure 5-1, is also a powerful means of communication, because the audience not only hears what you are saying, but also sees what you are saying.

Because audiences notice movements of your feet and hands, you should be particularly aware of those movements. Many movements of hands and feet by less experienced presenters do not contribute to the effectiveness of the presentation. One common example is playing with a tie, necklace, or belt. In general, you should avoid repetitive movements such as opening and closing a pointer, dancing from one foot to the other, or pac-

Figure 5-1. Karen Thole making a presentation of her fillet design for vanes in gas turbine engines.[12]

ing from one spot to another like a caged lion. These movements have a hypnotic effect, much like a hypnotist's watch, on the audience.

Other types of movement to be careful about are movements involving projection equipment. As mentioned in Chapter 4, you should practice with projection equipment so that you can efficiently turn on the equipment, change slides, and incorporate films and demonstrations. There are many stories about presenters who could not correctly place a transparency onto an overhead projector. Equally important with turning on the projector is turning off the projector. Many speakers make the mistake of leaving overhead projectors on when nothing but a bright white light is shown on the screen. Turning off the projector not only eliminates the light, but also cuts out the sound of the projector's fan, reducing the noise in the room and making it easier for the speaker and the audience to hear the questions.

In addition to turning on and off the equipment and

changing slides, another movement associated with a projected slide is to point out features on the projected image. A long metal or wooden pointer works well, because the speaker can point out the feature and stay out of the projector's light, which can be blinding. Moreover, a long metal or wooden pointer allows the speaker to ground the pointer against the screen. A common mistake made in pointing to a projection from an overhead projector is for the speaker to point directly at the transparency rather than at the screen. If the speaker's pen or finger quivers, that quivering becomes amplified by the projector.

Besides wooden or metal pointers, another common pointer is a laser pointer. Be careful with where you aim laser pointers; more than once I have ducked beneath a laser beam that a presenter had inadvertently aimed at the audience. Also be careful with how steady you hold a laser pointer, because laser pointers amplify a person's nervous movements. A slight quiver of the hand becomes amplified into a wild vibration on the screen. A speaker with a nervous hand should try anchoring the laser pointer against his or her side.

Eye Contact. A different type of movement to be careful about involves your eyes. Your eyes affect the audience. If you look at the floor, the audience will look at the floor. If you stare out the window, your audience will stare out the window. If you engage the audience with your eyes, the audience will return the look and will concentrate more on what you have to say. Granted, part of that increased concentration arises from guilt. When you are looking at an audience member, the person thinks, "I better pay attention because this speaker is looking at me." Another part of the increased concentration, though, arises because the audience member feels a part of the presentation.

How much should you look at the audience? Much advice exists in books about the number of seconds that you should look at someone. Rather than becoming self-conscious about that, you should just make sure that before the presentation is over you have made eye contact with everyone in the room if the audience is small. If the audience is large, make sure that before the presentation is over you have looked several times at every section of the room and that you have made eye contact with individuals in those sections. One myth about eye contact is that you should look above the heads of people to the wall in the back. Such a strategy makes no sense at all. With eye contact, you are both trying to engage the people in the room and to discern how they are responding to what you have to say. What could anyone possibly learn from looking at the wall?

Paying Attention to the Audience

For a presentation that she gave, the physicist Lise Meitner described her interaction with the audience in the following way: "[I] spoke loudly and looked at the audience and not the blackboard, although under the circumstances the blackboard seemed far more appealing than some of the people."[13] Looking at the audience is important, because even when they are not asking questions, your audience communicates to you. They speak to you with their eyes. When they stare intently at you, they tell you that they are concentrating on your message. When they nod their heads, they indicate agreement with your message. When they close their eyes or stare at the floor, they tell you that they have probably quit concentrating.

Many in the audience also speak to you with their facial expressions. Although some audience members keep a straight face through the entire presentation, others reveal if they are delighted, confused, angered, or bored. A good speaker pays careful attention to the audience and adjusts the presentation to engage the audience again if they begin to drift off. Such changes might involve slowing the pace if the audience is confused, speeding the pace if the audience is bored, or deleting tangential points if the audience is tired.

Although you should be sensitive to the mood of the entire audience, you should not overreact to the reaction of any one individual. For instance, you might encounter an audience member whose countenance is so angry that it frightens you to the point of distraction. In such cases, it is best not to look directly at that person. Perhaps that person has had an awful day and the expression of anger is not for you, but for someone else. Other times, you might have an audience member who is going to fall asleep on you no matter how well you present. In such cases, let the person sleep and focus on the rest of the audience. Perhaps that person has a new baby, and for that person your presentation is going to be the only quiet hour of the day.

During the asking of a question, the audience speaks to you directly, and your most important task is to listen. Such a statement might seem obvious, but more than once I have pushed through a difficult presentation, taken a deep sigh when I concluded speaking about my last slide, and then completely missed what the first questioner asked me. In those situations, the best I could do was to politely ask whether the person could repeat the question.

In his first presentation, Feynman made this same mistake and regretted it years later, because Wolfgang Pauli had apparently made a comment as to why Feynman's theory was incorrect. Feynman believed that had he listened, then perhaps he could have corrected the theory.[14] Feynman's experience points out one of the values of taking questions: the opportunity to receive feedback on our work from the audience. Although often a source of fear for presenters, question periods are opportunities to gain insights into the work from colleagues who are looking at that work with fresh eyes.

Although the issue of handling questions is discussed in more detail in Critical Error 10, given here is one example of how a panel of three speakers failed to pay attention to the audience during a question period. These three speakers had just given an interesting discussion about a scientific topic to a crowded room. Each speaker had spoken for five minutes, and then the three had discussed four prearranged questions for another fifteen minutes. The floor was then opened to questions for an additional fifteen minutes. The first questioner stood and began speaking—*rambling* would be a more accurate word. The person continued rambling for another ten minutes. Perhaps the person went on even longer, but several in the audience (including me) became so disgusted that we left. At first, we were disgusted with the questioner, but after a couple of minutes, our anger turned toward the speakers. They sat confused on the stage and continued waiting for the question to end. The questioner had no intention of ending the question because the questioner had no question. All that questioner had was a desire to talk.

After one minute, it was clear to most of the audience that this questioner had no question. Many of us in the audience gave signals to the speakers on stage: We looked away from the questioner and glanced down at our watches. The speakers on the stage did not pick up on our impatience. Later, many of us in the audience became even more demonstrative by speaking to one another, letting out our breaths in disgust, and raising our watches in front of their eyes. Still, the speakers remained passive. The audience had done everything that it could do to motivate the speakers to seize control of the presentation. Now the speakers had to act. That they did not was a mistake on their part, and their failure to seize control undermined what had been a worthwhile presentation.

The point of this example is that a strong scientific presentation is a two-way form of communication. The speaker certainly has the primary role of preparing, organizing, and presenting the information; however, the speaker has a responsibility to learn from the audience and to adjust the presentation to reach that audience.

Paying Attention to the Time

In 1841, William Henry Harrison gave a two-hour presidential inaugural address in a freezing rainstorm. Shortly after the address, he caught a cold and developed pneumonia. Two months later he died.[15] Most likely, catching pneumonia will not be a speaker's penalty for going too long, but taking too much time can have serious consequences.

For instance, speaking for too long damaged the reputation of an engineer at a recent international conference. The conference had designated a memorial session to honor the work of an engineering researcher who had unexpectedly passed away during the past year. The memorial session brought in more than two hundred audience members. One of those audience members, seated in the front row, was the researcher's widow. With five presentations scheduled, the memorial session was to reflect upon the work of the researcher. The first presenter was a colleague who had worked with the researcher very early in his career, and the fifth presenter was a colleague who had been working with the researcher at the time of his death. For this session, the presenters were to show the work of this researcher as it had progressed over three decades and to emphasize this researcher's large contribution to the field. Everything went smoothly until the last speaker's presentation. This speaker, like the others, had twenty minutes to make the presentation, but unlike the previous speakers, this speaker did not stop his presentation after twenty minutes.

The session chair quietly signaled the speaker that the end had arrived, but to no avail. The speaker continued speaking through the five-minute break that had been scheduled between this memorial session and the concurrent sessions that were to follow. Meanwhile, the audience became restless. Several in the audience were scheduled to give talks at the concurrent sessions, and they wanted to head into their rooms and set up things. Because the session was a memorial session for a respected researcher, though, no one left.

After a few more painful minutes, the session chair stood and asked the speaker to end his presentation. Still, the stubborn speaker continued; he had a stack of slides and he was determined to get through them. People became increasingly uncomfortable. Exasperated, the session chair walked up and removed the slide currently being projected, but the hardheaded speaker put another one in its place. Finally, after this speaker had placed all his slides onto the projector, he ended his talk, and the session abruptly dispersed. The consensus afterwards was that this memorial session, which was to leave participants with the memory of work by a departed colleague, ended up leaving participants with an uncomfortable mixture of emotions: anger at the speaker who had not planned for the situation and embarrassment for the departed researcher's wife, who had to endure the awkward exchanges between the session chair and the bullheaded speaker.

So how do you stay within the time limit of a presentation? Like most questions raised in this book, the answer to this question depends upon the situation. In presentations in which the speaker is not interrupted by questions, the speaker has control over the time. In presentations, though, in which the audience can interject questions, the control of time is shared between the speaker and audience.

Consider first the situation in which the presenter is allowed to speak without interruption from the audience. For this situation, the most important step in making the deadline is preparation. This preparation includes defining a scope that you can cover adequately in the time allotted. It also includes planning to show no more slides than the time allows. For instance, planning to show twenty slides for a fifteen-minute talk makes no sense, because the audience needs what my colleague Harry Robertshaw calls "soak time" to process each slide. Even a title slide such as that shown in Figure 5-2 should be allotted at least sixty seconds, not only so that the audience can become oriented to the topic, but also so that the audience can become accustomed to the speaker's delivery. Moreover, if a slide includes a complex graphic (see Figure 5-3), even more time is needed.

Another important aspect of preparation is to practice the presentation all the way through with the slides. That practice is important not only for giving yourself the confidence that you can find the words to explain each idea, but also to develop some confidence that your presentation will stay within the time limit. Certain variables cause the time achieved during the rehearsal to differ from the time achieved in the actual presentation. One variable is the effect of nervousness. Although most people, when nervous, will speak more quickly before a live audience than they will speak alone in a hotel room, a few people actually speak more slowly. If you are that second type of person, then you have to factor in that difference as well.

Another variable is digression. Some people, including myself, are inclined to add a story or dwell too long

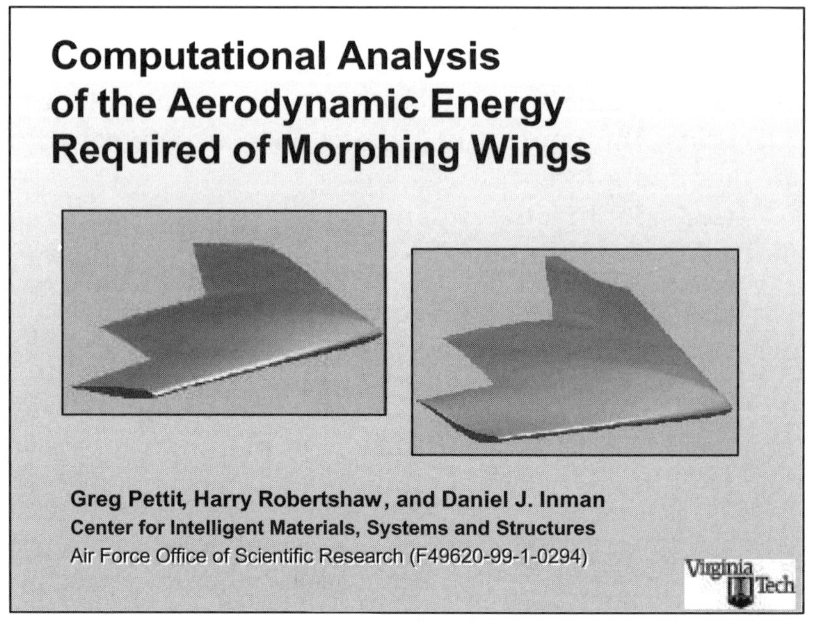

Figure 5-2. Example title slide.[16] Even with a title slide, the presenter should allot at least sixty seconds for the audience to become oriented to the topic and to become accustomed to the presenter's delivery.

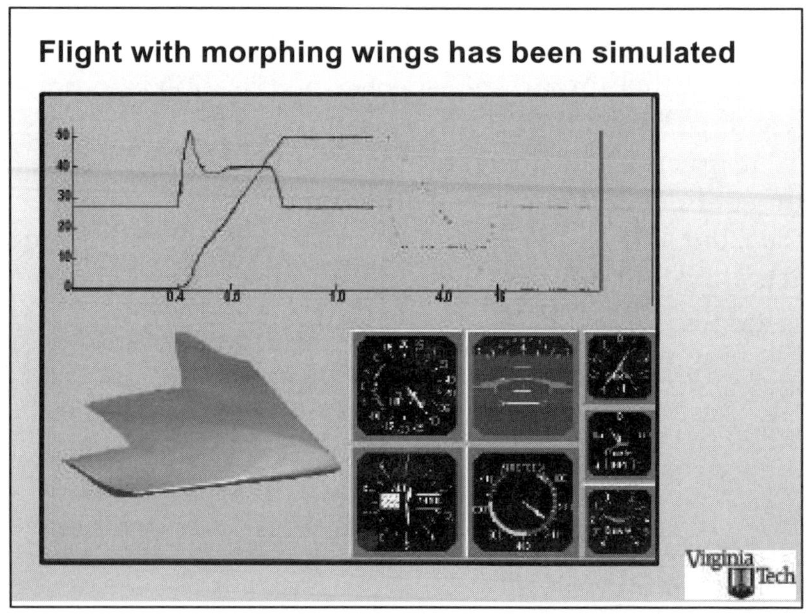

Figure 5-3. Example slide that has a complex graphic.[17] A presenter could easily spend 2 or 3 minutes on this slide.

on an interesting point, especially if the audience is responding positively to the material. For those of us who have this habit, having a benchmark during the presentation is important. One benchmark might be finishing the introduction by a certain time. Another might be beginning the second main point of the presentation at a certain time. A quick glance of your watch as you change slides allows you to check your progress. Another tip that works well when one is using benchmarks is to have an optional point to cover. If you are on time at your benchmark, then you would cover that point; if you are running behind, you would skip it.

When the occasion is such that the audience can interrupt you, meeting the time limit is more difficult. In essence, the audience becomes part of the presentation. Although they are players, you are still the leader, and, if appropriate, you should exercise your authority to keep things moving. One way to do that is to postpone questions that are premature, in other words, questions that will be addressed later in the presentation. Another way to keep things moving is not to address questions that are outside the scope of the presentation. In such cases, offer to speak with the individual after the presentation and then get the presentation back on track. If you are diplomatic in your responses to such questions, most questioners will accept your decision. Some people, however, are not so respectful. In such cases, you have to read the situation. If the person is your manager, you will probably have to allow the manager to have his or her time on that issue. Also, if several people in the audience want to discuss a tangential issue, it might make sense to defer. However, if only one person is holding back the others, you should keep things moving. Otherwise, the audience's irritation at the one pressing the question might turn to anger at you for not keeping the presentation on track.

Critical Error 10
Losing Composure

On Monday and Wednesday, my mother was nervous and agitated from the time she got up. At five o'clock on these days she lectured. After lunch she shut herself into her study in the Quai de Béthune, prepared the lesson, and wrote the heads of chapters of her lecture on a piece of white paper. Towards half-past four she would go to the laboratory and isolate herself in a little rest room. She was tense, anxious, unapproachable. Marie had been teaching for twenty-five years; yet every time she had to appear in the little amphitheater before twenty or thirty pupils who rose in unison at her entrance she unquestionably had "stage fright."[1]

—Eve Curie

To watch a presenter lose composure is a painful experience. In some cases, the presenter has so much stage fright that he or she loses composure before the presentation even begins. For instance, a graduate student of one my colleagues lost composure at a conference recently. This graduate student, who had to present two papers back to back, was so nervous that he simply read the papers rather than look the audience members in the eyes and speak to them about what he knew. Moreover, he read so quickly that he finished both papers in less than the time that was allotted for just one. Needless to say, no one in the audience learned anything from the presentations.

In other cases, the loss of composure arises during the presentation because something unexpected occurs. More often than not, these cases occur during questioning. A loss of composure during the question period often follows a certain sequence. A question trips up a speaker, and the speaker loses confidence. Some members in the audience sense a flaw in the work and ask more questions on the same point. The speaker loses even more confidence, and the questions become more biting.

Controlling Nervousness

Just the act of speaking before an audience generates much nervous energy for presenters. There is nothing unusual about feeling nervous before a presentation. Many great scientists such as Marie Curie and Richard Feynman have shared these same feelings. What is most important is that the nervousness not pull down the presentation or inhibit the speaker from making presentations that he or she should.

Unfortunately, shyness has inhibited a number of excellent scientists and engineers from making presentations. The chemist Fritz Strassmann, for instance, allowed Otto Hahn to make the presentations of their work on nuclear fission. Interestingly, for their work, Otto Hahn (and Otto Hahn alone) won the Nobel Prize. Many believe that Strassmann should have shared in that prize.[2] Moreover, many more feel that Lise Meitner, who shied away from making presentations in the early years of her career, should have shared in the award. According to Ruth Sime, Meitner "initiated the experiment, and with Otto Frisch explained the [fission] process."[3] At the time of the discovery, Meitner, who was of Jewish heritage, was not in Germany with Hahn and Strassmann. She had fled to Sweden to escape persecution from the Nazis.

Another shy scientist was Robert Corey. Corey, who collaborated with Linus Pauling, allowed Pauling to make the presentations and in so doing allowed Pauling to receive the lion's share of the credit.[4] Yet another scientist who was shy before crowds was the Nobel winner Christiane Nüsslein-Volhard, who sometimes had her graduate students give presentations that she should have made herself.[5]

Overcoming your nervousness to the point of simply making the presentation is not enough. What distinguishes the best speakers is their ability to channel that nervous energy into positive energy that serves their presentations. When you do not channel the nervous energy in a positive manner, it often comes out in distracting movements: jingling coins in a pocket, playing with a pointer, dancing a samba with one's feet.

Nervousness also affects the voice. For a talk that she had spent many hours preparing, Christiane Nüsslein-Volhard was so nervous that her voice shook for several minutes.[6] Nervousness affected Richard Feynman's voice in a different way. For the second talk that he ever gave, a ten-minute presentation at the American Physical Society Meeting in New York in February 1941, Feynman lost the nerve to speak to the audience and simply read his speech in what he termed a fashion that was "dull" and "impossible for people to understand."[7] In yet another example, the nervousness of Luis Glaser, who is now provost at the University of Miami, caused him to speak much too quickly for a seminar talk that he gave as a graduate student. The talk was for his research group, which was run by Nobel winners Gerty and Carl Cori. Although the talk was scheduled to last fifty minutes, Glaser rushed through it in thirty. To teach him a lesson, Gerty Cori had everyone remain in the room, essentially in silence, until the fifty minutes was up.[8]

So how do you make nervousness work for you? First, you should think positive thoughts. The nervousness that a presenter feels is similar to the nervousness that an athlete feels. How do athletes handle nervousness? Many successful tennis players imagine success. Steffi Graf, for instance, believes that positive thinking is a powerful force in playing tennis.[9] So does Jimmy

Connors. Connors, while waiting to return a service, imagines not only hitting the ball but also the flight of his service return across the net. In his book *The Inner Game of Tennis*,[10] Tim Galloway presents an excellent discussion of the power of positive thinking in one's tennis game.

Many successful basketball players also imagine success, particularly in their shooting. One player whom I often see is Ieva Kublina, the power forward for Virginia Tech. Like many players, Kublina has a beautiful shot: excellent form and a nice arc. What distinguishes Kublina, though, is her height (she is six-foot four), her range (she is accurate even from the three-point line), and her conviction that the shot will go in. Some players, after missing a couple of shots, will begin to hesitate on their shots and allow nerves to alter their shooting form. Not Kublina. She continues to go up strong with the same form and the same belief that the ball will go in. If the team badly needs a basket, Kublina is the person that the coaches call upon to shoot. She wants to shoot the ball, she believes that it will go in, and she has the talent to match her conviction.

If you are combatting nervousness before a presentation, consider adopting the positive attitude of a successful basketball or tennis player. Imagine yourself delivering a successful presentation. Imagine yourself delivering each of your main points. Imagine the audience focused on your message and nodding in agreement.

Sometimes a bad case of nervousness means that the speaker is not prepared for the presentation. Before an important presentation, you should have two or three practice runs. On at least one practice run, you should incorporate your visual aids. If you are unsure about the presentation, have a colleague or two attend. These critiques should occur such that you have enough time to incorporate valid criticisms. If you try to make major changes right before a presentation, you might end up causing more harm than good.

If you have done your preparation, then the structure, speech, and visual aids of your presentation should be ready. What is left are the little things that take place in the presentation room to smooth your delivery: making sure that the projector is focused, checking the order of your overheads, and adjusting the lights. Once you have set things up in the room, you should concentrate on your listeners. Meet them before the presentation and ask them questions. By concentrating on your listeners, you shift your thoughts, and worries, away from yourself and give needed attention to your audience. Remember: You are working for them. If you can focus your attention onto your audience, then any residual nervous energy is going to work for the presentation, not against it.

Another way to overcome nervousness is to understand its cycle. David Bogard, a mechanical engineering professor at the University of Texas, claims that each of his bouts with nervousness subsides as soon as he begins the presentation.[11] Bogard says that knowing that the churning and wrenching of his stomach will end is a comfort to him in the days leading up to the presentation. Richard Feynman made the same claim. For his first presentation, Richard Feynman faced an intimidating audience: the mathematician John von Neumann, the astronomer Henry Norris Russell, and the physicists Wolfgang Pauli, Eugene Wigner, and Albert Einstein. Feynman remembered how nervous he was before that presentation.[12] His hands shook noticeably in removing his notes from the envelope. What Feynman also recalled was that the nervousness subsided as soon as he began the presentation and concentrated on the subject.

Once, when standing backstage with a nervous presenter, Mark Twain said, "Don't worry—they don't expect much."[13] As much as any piece of advice I have received, Twain's advice has helped me change my attitude about nervousness. Once I see the tentative looks of the audience filing in, I realize that most of them are resigned to yet another boring presentation. At that moment, the presentation becomes a challenge in the positive sense. To myself I say, "They think that I am going to bore them. Well, I'll show them."

In addition to fighting nervousness before the presentation, speakers sometimes have to battle nervousness during the presentation, especially when things do not go as expected. About two-thirds of the way through one of my first presentations, which was to an audience of about one hundred, about half of the audience got up and walked out. Up to that moment, I had been feeling positively about this presentation. Everyone had been attentive and there were no signs of boredom. However, all those people walking out crushed me on the inside. I felt like throwing up my hands and quitting, but I focused on those who remained in the room and finished, trying to act as though nothing had occurred. What I learned later was that those people who walked out had to attend a required meeting. From that incident I learned an important lesson: "No matter how bleak things look, do not lose your cool."

Distractions often occur in presentations. Light bulbs for projectors periodically go out. Fire alarms occasionally go off. People in the audience sometimes have to leave, sometimes cannot stay awake, sometimes stare absentmindedly at your shoes, sometimes talk with another, and sometimes are so preoccupied with personal problems that they wear scorns on their faces. The first time that each of these incidents occurred in one of my presentations, my stomach started churning. Although I kept my cool on the outside, the experience was wrenching. Looking back, though, I see that none of those incidents were that important. The audience did not hold me responsible for the light bulbs, the fire alarms, or the reactions of other audience members—only for the way that I reacted to those incidents.

Another time that he had to give a seminar for the Coris' research group, Luis Glaser had a surprise waiting for him. Just before the talk, Gerty Cori asked him what the subject of his talk was. When he told her, she said, "That bores me." Glaser's reaction was to go on up and give his talk as best he could. As he reasoned, "What else could I do?"[14] I admire that response. In the face of such devastating criticism from such an admired figure, another presenter would have folded, but Glaser went on and did the best that he could with what he had. In the end, that is really all each of us can do.

Handling Questions (Even the Tough Ones)

How should one handle questions? As mentioned in Critical Error 8, the first step in handling a question is to listen to the question. That step might seem obvious, but after finishing the formal presentation part of the talk, many presenters relax, forgetting that the scientific presentation is not over.

If you do not understand the question, you should not hesitate to ask for clarification. After all, the question is something that the audience member has probably just come up with and not rehearsed. Once you understand the question, you should repeat or rephrase it if the room is so large that the rest of the audience has not heard it. You should also think before answering. A pause is justified and often appreciated by the audience. If you know an answer to the question, you should then answer that question, but make sure to balance two concerns: satisfying the questioner and doing so concisely so that others have a chance to ask questions.

What if you do not know the answer to a question? Many people fear receiving a question that they cannot answer. Much about this fear is unfounded. For one thing, the audience does not expect you to know everything about the topic. For another thing, many questions concern topics outside the scope of the presentation. If you do not know the answer to a question, you should think about whether that question actually lies within the scope of the presentation. If not, then you should state that.

If the question does lie within the scope of the presentation and you do not know the answer, you should *not* try to bluff an answer. If you are exposed (and the chances are high that you will be), your credibility will quickly sink. Worse yet, the sharks in the audience will smell blood and begin to circle. If you do not have a complete answer, you should admit that you do not have a complete answer, but then state what you do know about the point questioned. In some cases, the actual answer might be something that no one knows. If that is the case and if you know that no one knows the answer, stating as much might win you respect. At the least, such an answer would show that you know the subject's literature. If the question is something that you should know, but have forgotten, you should promise the questioner that you will look up the answer after the presentation; and then you should do so.

One of the most difficult situations occurs when a questioner challenges you. In many cases, the purpose is not a personal attack. Many great scientists such as Wolfgang Pauli,[15] Rosalyn Yalow,[16] and Gerty Cori[17] rigorously challenged work that they felt was inaccurate. Although such challenges often strengthen the science, these challenges also overwhelm many presenters. What should you do in such a situation? My advisor, Kamalaksha Das Gupta, who studied under the great Bose, used to tell us that whenever someone challenged our work in the question period, we should stand very straight and answer in a loud voice for everyone to hear. Das Gupta said that even if all we knew to say about that point was just what we had said in the formal part of the presentation, we should say it loudly and confidently.

Another strategy is to fight back. On her comprehensive examination at the University of Illinois, Rosalyn Yalow came under attack from the department chairman. After she had solved the examination problem that he had posed, he asked her to solve the problem a different way. She refused, saying that Goldhaber and Nye (two faculty members in the department) had taught her this way and that if there was anything wrong with that method, then he should speak with them about it. The chairman walked out of her exam and did not return.[18]

In her first scientific presentation, the Nobel winner Gertrude Elion also stood her ground when a distinguished researcher questioned her conclusions. Holding one's ground does not mean that animosity has to develop between the speaker and the questioner. In Elion's case, for example, immediately after the presentation the researcher invited her to lunch, where she had the opportunity to explain her work in depth.

In one of his first presentations, David Bogard, from the University of Texas, was challenged on the assumptions of his work. Because the questioner's voice had a sarcastic tone, Bogard felt that the questioner was going after him, perhaps because Bogard was new in the field. Such a challenge demanded a strong response, because if the audience were to consider the assumptions flawed, then they would have considered the work worthless. Fortunately, Bogard had done his homework on the literature. Knowing that he had the goods on this question, Bogard calmly placed a foot on a chair and began counting his reasons for making his assumptions. First, he recalled one paper in the literature that supported his assumptions. Then he recalled a second, and then a third and a fourth and a fifth. By the end, two things were clear to the audience: Bogard had read the literature, and the sarcastic questioner had not.[19]

To see the sharks circle as a speaker loses confidence is a sad thing. More than once I have felt the waters begin to churn and I have had the flashing thought that I am losing control of my presentation. In such cases, my instinct has been to do what Das Gupta advised: stand up straight, raise my voice, and repeat my strongest evidence for the assertion.

When attacked by a harsh question, Ronald Reagan took a different tack. In such cases, he lowered his voice rather than raising it. His voice adopted that grandfatherly sound. In lowering his voice this way, Reagan guided the sympathies of the audience to his side. The audience subconsciously thought, "Why is that questioner being so mean to that old man?" Lowering your voice can be effective as long as you remain resolute. For example, Marie Curie spoke softly, but resolutely.[20] What you do not want to do is to stumble with filler phrases such as "um" and "uh." Those make you appear weak and confused.

John F. Kennedy shows yet another strategy to handle attacking questions. When questioned harshly about whether it was ethical for him to have named his own brother as attorney general, Kennedy paused and fixed his eyes on the questioner. Then, he suddenly said no, turned, and called upon another questioner in a different part of the room. In acting so decisively, Kennedy did not give the original questioner a chance to follow up.

What do you do if a questioner attacks your work and you realize that the questioner is correct? Einstein and Bohr provide us with courageous examples of what we should do, but what few of us would dare. After a presentation, Einstein fielded a question from a young, unknown Russian whose broken German conveyed something along the lines that what Einstein had said "was not so stupid."[21] The Russian turned out to be Lev Landau, who became one of the Soviet Union's greatest theoretical physicists. In the question, Landau pointed out an error in one of Einstein's equations. While everyone in the room was chastising Landau for his rudeness, Einstein studied the blackboard and thought about what Landau had said. Finally, Einstein turned back to the audience and quietly said that the point that the young man had raised was correct and that what had been presented today beyond a certain step was incorrect. This statement reveals that Einstein's quest was not personal glory, but the search for truth. Niels Bohr had similar aims and never hesitated to admit when he was in error,[22] a trait in his character that earned him much admiration. Perhaps that is the best sign of one's security: the willingness to admit when one is wrong.

Chapter 6

Conclusion

[The lecturing of Boltzmann] was the most beautiful and stimulating thing I have ever heard.... He was so enthusiastic about everything he taught us that one left every lecture with the feeling that a completely new and wonderful world had been revealed.[1]

—Lise Meitner

In their careers, scientists and engineers make many presentations, and these presentations are expensive in terms of the time expended by the audience to attend and by the presenter to prepare. Yet all too often, scientific presentations are not nearly as effective as they could be at either communicating the information or persuading the audience. All too often, the presenter creates a presentation without contemplating the situation: the audience, the purpose, and the occasion. For instance, one common error is presenting the information at too complex a level for the audience to comprehend. Another common error is relying on defaults from presentation programs such as Microsoft's PowerPoint to produce presentation slides that are burdened with bullets and void of needed images.

In regard to presentation slides, this book has called for a revamping of the designs used by most scientists and engineers. Instead of recommending slide designs that rely on phrase headlines and bulleted lists, this book has called for presentation slides anchored with sentence headlines and supporting images. Although such slides take more time to create, the benefits in a scientific presentation are more than worth the extra time expended. One benefit is a better orienting of the audience during the presentation. A second benefit is a better orienting of the audience after the presentation, when the slides are used as notes. Perhaps the most important benefit, though, is that adopting this design forces the presenter to decide whether each slide is necessary.

Although this book has devoted much space to the design of presentation slides, the essential ingredients for a strong presentation are more basic. For a strong presentation—a presentation that not only delivers the information, but that truly engages the audience—three ingredients have to be present. First, the speaker must understand the subject. The speaker is not expected to know everything about the subject, but what the speaker imparts has to be worth the audience's time. A second essential ingredient is that the speaker must have a keen awareness of the audience: what they know about the subject and why they have attended. The third essential ingredient is that the speaker show a genuine enthusiasm for the subject. Not every speaker has to present with the passion of Linus Pauling or Richard Feynman, but every speaker should instill in the audience a respect for the subject.

A great presentation is remembered for a long time. Decades later, Lise Meitner claimed that she could remember every detail from the first lecture that she heard Einstein give.[2] At that lecture, Einstein explained that energy is trapped in mass, according to the now famous equation $E=mc^2$. This book has highlighted what distinguished the presentations of Albert Einstein and other model presenters: Ludwig Boltzmann, Richard Feynman, Rita Levi-Montalcini, Linus Pauling, and Chiung-Shien Wu. The presentations of these individuals touched many people and, as Meitner pointed out, have had long-lasting effects.

If there were one piece of advice about presentations that I could whisper into the ears of every scientist and engineer, it would be to aim higher. In other words, do not be content to present in the staid fashion to which so many presenters resign themselves. Rather, for your audience, purpose, and occasion, you should strive to craft a presentation that is truly worth your audience's time, a presentation that your audience will not forget.

Appendix A

Checklist for Scientific Presentations

Table A-1. Checklist for scientific presentations.*

Speech

Necessary information conveyed?	Assertions supported?
Audience targeted?	Tone controlled?
Terms defined?	Examples given?

Structure

Organization of Beginning	Transitions
Scope defined?	Beginning→middle?
Topic justified?	Between main points of middle?
Proper background given?	Middle→ending?
Talk memorably mapped?	
Organization of Middle	Emphasis
Divisions of middle logical?	Repetition used effectively?
Arguments methodically made?	Placement used effectively?
Organization of Conclusion	
Main points summarized?	
Closure achieved?	

Presentation Slides

Slides orient the audience?	Slides show key images?
Slides are clear to read?	Slides show key results?
Slides have proper level of detail?	Slides show talk's organization?

Delivery

Speaker controls nervousness?	Eye contact made?
Speaker shows energy?	Movements contribute?
Speaker exudes confidence?	Equipment handled smoothly?
Voice engages?	Questions handled convincingly?
Speed is appropriate?	Questions handled succinctly?
Filler phrases ("uh") are avoided?	Time is appropriate?

*Not every item on this list applies to every presentation.

Appendix B

Design of Scientific Posters

Posters are a special type of presentation. When well designed, posters are not simply journal papers pasted onto boards. Nor are they mounted sets of PowerPoint slides. Rather, posters, when effectively designed, are a medium distinct in typography, layout, and style.

The purpose of a poster is to present work to an audience that is passing through a hallway or exhibit. When posters are displayed at conferences such as the display depicted in Figure B-1, the presenter usually stands next to the poster. This arrangement allows for passersby to engage in one-on-one discussions with the presenter. When posters are displayed in the hallways of laboratories, universities, and corporations, the posters typically stand by themselves.

Figure B-1. Poster presentation of capstone design projects at Pennsylvania State University.[1]

For a poster to communicate the work, the poster first has to orient an audience that is not seated at a desk, but that is moving through a hallway or auditorium. Often the audience has distractions of noise and movement from other people. Given those distractions, a journal article tacked onto a board fails as an effective poster, because the audience cannot concentrate for a long enough time to read through the paper. In fact, given the distractions that the audience faces, many in the audience will not even bother trying to read a journal article presented in that fashion.

Because the audiences for posters vary so much, coming up with specific design rules for posters is difficult. For instance, some posters target general audiences and therefore focus on the results and importance of the work. Other posters target specialists in the field. For such posters, the audience expects to learn not only what the results are, but how they were achieved. Still other posters are intended for a mixture of these audiences.

Another reason that coming up with guidelines is difficult is that posters are created in significantly different ways. From the audience's perspective, the best way is as a single large sheet of paper. To produce such a poster, you need a special printer. The advantages of this method are that the poster has unity, looks professional, and can be quickly mounted. The disadvantages of this method are that the printing costs are relatively high, revisions require additional printings (and therefore even higher costs), and transportation requires a long tube that can be cumbersome to transport.

A second way to create a poster is to print out pieces of the poster and then to assemble those pieces onto a single sheet or board. The advantages of this method are that the printing costs are lower and that changes can be made more easily. The disadvantages are that the poster does not look as handsome as the single printed sheet and that assembly can take much time.

Yet a third way to create a poster is to modify and arrange a set of presentation slides printed on paper such that the set can be read as a poster. This third way is usually the easiest to create, because most scientists and engineers already have sets of presentation slides about their projects. For this type of poster to succeed, though, the presenter has to design the slides so that the audience can follow the work. Slides that rely on phrase headlines and that lack key images generally fail. Slides that follow the principles given in Table 4-2 do much better; still, the presenter has to arrange those slides so that the audience reads them in the appropriate sequence. Otherwise, the audience does not see the organization.

Table B-1 presents guidelines for the design of posters. Although these guidelines are for the single-sheet poster, the principles apply to the other two types. As with the guidelines for slides, these guidelines are divided into guidelines for typography, layout, and style.

Much about the typography of slides applies to posters. The typeface should be thick enough to be read from a distance. For that reason, a uniform font such as Arial is appropriate. The presenter should avoid a font such as Garamond that has thin strokes. Also, because the title is read first, it should be the largest block of type on the poster and should be centered or placed in the upper left-hand corner. Moreover, because readers use headings to find particular information, the presenter should boldface the headings so that they stand out. As in designing presentation slides, the presenter should avoid blocks of capital letters in the design of posters.

In the layout of a poster, the presenter should arrange the sections and graphics such that the order of reading is clear. A common confusion is whether to read down or across. One way to indicate that the audience is to read down rather than across is to make the gutter (the distance between the columns) wider than the vertical

Table B-1. Guidelines for designing posters.

Typography
Use a typeface such as Arial that is thick enough to read
Boldface the title and headings
Use type sizes of 18 points or higher (14 points okay for references and footnotes)
Avoid blocks of all capital letters
Layout
Arrange sections such that the order of what to read is clear
Be generous with white space
Keep lists to two, three, or four items
Keep text blocks to just a few lines
Style
Include an orienting image near the title or in the background
Opt for vertical lists rather than long paragraphs
Where possible, opt for graphical presentations rather than lists or paragraphs
Accept the fact that a poster cannot present as much detail as a journal article can

distance between sections. A second way is to place a graphic in the second column, as shown in Figure B-2.

A second guideline for the layout of posters is to include enough white space that key points are emphasized. Where text borders white space is where emphasis occurs. Note that a poster that has generous white space between sections is much more inviting to the reader. Conversely, a poster that is packed with text and graphics intimidates the audience. A third guideline for the layout of posters is that the presenter should limit lists to two, three, or four items. Likewise the blocks of text, either in listed items, in section paragraphs, or in figure captions, should be neither too long nor too wide. To determine these limits for what is too long and too wide, the presenter should mount a draft of the poster and read the poster as the audience will. If the presenter finds himself or herself wanting to skip a section, then that section is probably too long or too wide.

As regards the style of a poster, the presenter should first consider including an orienting image or set of images either near the title or as the poster's background. Such images reinforce for the audience what the poster is about.

A second stylistic point for posters is to consider carefully whether to use paragraphs, lists, or graphical representations for the different sections. Paragraphs, which are the primary means for presenting groups of ideas in journal articles, have the advantage of showing the connections between the individual ideas. For that reason, paragraphs are more effective at revealing the logic of arguments. A disadvantage of paragraphs, though, is that on a poster they are intimidating to read, especially when they are long. One way to circumvent that intimidation is to use a list of points, as in the posters of Figures B-2 and B-3. For a list to be effective, the number of items should be limited (two, three, or four items, if possible), and the length of any one item should be short (just a few lines).

Instead of paragraphs or lists, a better way to present information in a section is graphically, such as in a flow chart. Although graphics take much time to prepare, they can communicate more efficiently than paragraphs and more memorably than lists (with too many lists, the reading becomes tiresome, even hypnotic).

In general, because of the situation in which the audience reads a poster, a poster cannot communicate as much information as a journal article can. When designing a poster, the presenter should accept this constraint and limit the poster to the essential information. If too much information is included, the poster will overwhelm the audience and in many cases cause passersby to give up on that poster and move on to the next one.

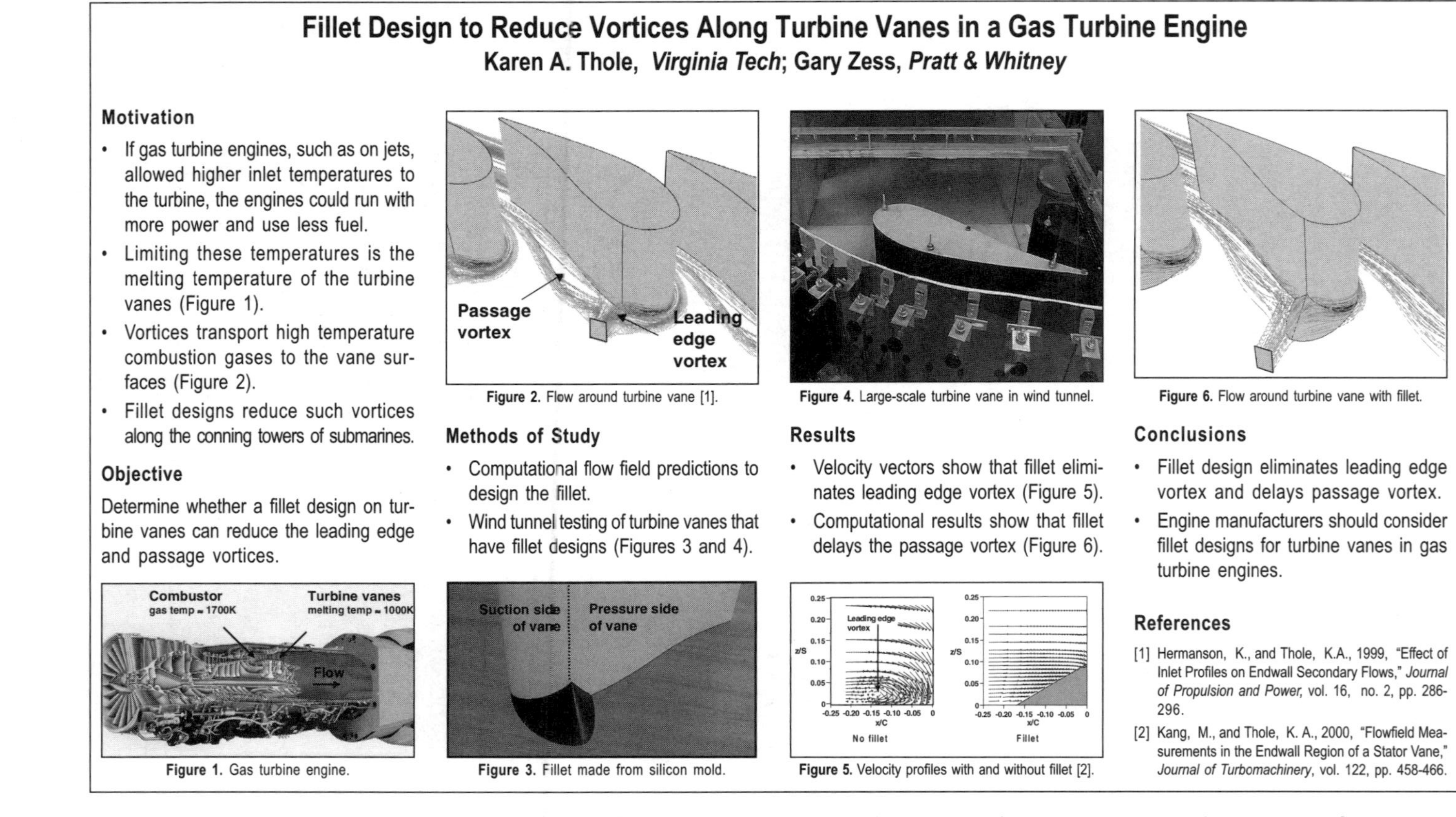

Figure B-2. Poster showing fillet design for improving the aerodynamics of vanes in a gas turbine engine.[2]

Gap-Crossing Decisions by Red Squirrels in Fragmented Forests

Victoria J. Bakker, *University of California, Davis*

Objective

To study factors influencing decisions by red squirrels (*Tamiasciurus hudsonicus*) to cross gaps in fragmented forests.

Forest-clearcut edge at study site, which is in the center of Mitkof Island and in the Tongass National Forest, Alaska. Logging is the primary land use.

Rationale

- Knowing how mammals move in fragmented forests can aid in location of reserves and corridors.
- Questions exist about which factors control decisions of mammals to cross gaps in their preferred habitats.

Translocation of individual squirrels across gaps for release and subsequent tracking.

Hypotheses

- Efforts to minimize predation risk, energy expenditures, or encounters with territorial conspecifics were hypothesized to control crossing decisions.
- Predation risk was assumed higher in clearcuts than in forests because of lower overstory cover and lack of trees for escape.
- Energy expended per distance traveled was assumed higher in clearcuts due to higher shrub stem densities.
- Conspecific encounter rates were lower in clearcuts than in forests.

Methods

- Documented home ranges and territorial behaviors of squirrels living near clearcuts less than 10 years old.
- Induced movement by translocating individuals across gaps and used radio-telemetry to document homing paths.
- Conducted call-back surveys along clearcut perimeters to determine conspecific defense levels.
- Used logistic regression to relate extrinsic factors, such as gap size, and intrinsic factors, such as body mass, to gap-crossing probability.

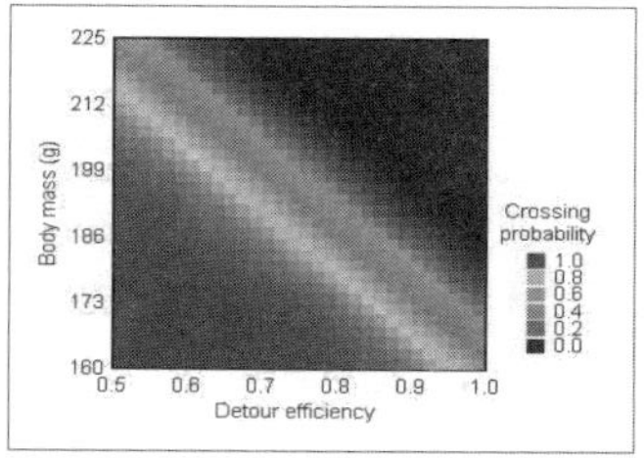

Determinants of gap crossing: Relationship between detour efficiency, body mass, and gap-crossing probability, based on logistic regression.

Results and Discussion

Of 30 squirrels translocated at 5 clearcuts, 11 crossed clearcuts and 19 detoured along forested routes.

1) Gap-crossing probability was inversely related to squirrel body mass and detour efficiency (η_D):

$$\eta_D = \frac{\text{Direct distance home}}{\text{Indirect distance home}}$$

2) Lighter squirrels were more likely to cross clearcuts. Squirrels in poor condition may take more risks when moving.
3) Squirrels were more likely to cross if detour efficiency was low, suggesting that squirrels assess distances of alternate routes and that predation risk, energetics, or both influence crossing decisions.
4) Squirrels choosing forested routes avoided the route with the greater number of highly defended territories.

Non-significant factors were crossing distance, clearcut size, clearcut age, and individual's territorial behavior.

Acknowledgments:
U.S. EPA STAR Program

Figure B-3. Poster to discuss results on research about the gap-crossing decisions of red squirrels.[3]

Notes

Preface

[1] Isaac Asimov, Foreword to *Linus Pauling: A Man and His Science*, Anthony Serafini (San Jose: toExcel, 2000), p. xiv.

[2] Michael White and John Gribbin, *Einstein: A Life in Science* (New York: Penguin, 1995), pp. 164–165.

[3] Ruth Sime, *Lise Meitner: A Life in Physics* (Berkeley: University of California Press, 1996), pp. 96–97.

[4] D.H. Frisch, private communication to Abraham Pais, "Reminiscences from the Postwar Years," *Niels Bohr: A Centenary Volume*, ed. by A.P. French and P.J. Kennedy (Cambridge, MA: Harvard University Press, 1985), p. 247.

Chapter 1

[1] Jagdish Mehra, *The Beat of a Different Drum* (Oxford: Clarendon Press, 1994), p. 482.

[2] Atul Kohli, engineer at United Technologies Corporation, Pratt & Whitney (4 December 2000), personal communication to author.

[3] Robert Pool, "Superconductor Credits Bypass Alabama," *Science*, vol. 241 (1988), pp. 655–657.

[4] Robert Pool, "Feud Flares over Thallium Superconductor," *Science* (2 March 1990), p. 1029.

[5] Patrick McMurtry, Professor of Mechnical Engineering at the University of Utah (March 1988), personal communication to author.

[6] Sharon Bertsch McGrayne, *Nobel Prize Women in Science* (Secaucus, NJ: Citadel Press Book, 1998), p. 37.

[7] Ibid., p. 236.

[8] Richard P. Feynman, *"Surely, You're Joking, Mr. Feynman!"* (New York: Norton & Company, 1985), p. 166.

[9] Michael Faraday, letter to Benjamin Abbott on 11 June 1813, *The Selected Correspondence of Michael Faraday*, ed. by L.P. Williams, R. Fitzgerald, and O. Stallybrass (Cambridge: Cambridge, 1971), pp. 60–61.

[10] Sharon Bertsch McGrayne, *Nobel Prize Women in Science* (Secaucus, NJ: Citadel Press Book, 1998), p. 181.

[11] Muriel Rukeyser, *Willard Gibbs* (New York: Doubleday 1942), p. 320.

[12] James D. Watson, *The Double Helix*, critical edition (New York: W.W. Norton & Co., 1980), pp. 43, 45.

[13] Ibid., p. 14.

Chapter 2

[1] Kelves, Daniel J., *The Physicists* (New York: Knopf, 1978), p. 218.

[2] Peter Goodchild, *J. Robert Oppenheimer* (Boston: Houghton Mifflin Company, 1981), p. 25.

[3] Nuel Pharr Davis, *Lawrence and Oppenheimer* (New York: Simon & Schuster, 1968), p. 27.

[4] Kelves, Daniel J., *The Physicists* (New York: Knopf, 1978), p. 218.

[5] Peter Goodchild, *J. Robert Oppenheimer* (Boston: Houghton Mifflin Company, 1981), p. 25.

[6] Otto Frisch, *What Little I Remember* (Cambridge: Cambridge University Press, 1996), p. 57.

[7] Ibid., p. 63.

[8] Fred Soechting, Engineer for Pratt & Whitney (Madison, WI: University of Wisconsin, 1996), presentation.

[9] Richard P. Feynman, *"Surely, You're Joking, Mr. Feynman!"* (New York: Norton & Company, 1985), pp. 244–245.

[10] Ellen Ochoa, NASA Astronaut, "The Atlas-3 Mission of Space Shuttle," presentation (Madison, WI: University of Wisconsin, 11 September 1996).

[11] Patricia N. Smith, Director at Sandia National Laboratories (Livermore, CA: November 18, 2000), interview with author.

[12] Geoffrey Cantor, *Michael Faraday: Sandemanian and Scientist* (New York: St. Martin's Press, 1991), pp. 151–152.

[13] Lise Meitner, "Looking Back," *Bulletin of the Atomic Scientists*, vol. 20 (November 1964), pp. 2–7.

[14] Carlo Cercignani, *Ludwig Boltzmann: The Man Who Trusted Atoms* (Oxford: Oxford University Press, 1998), pp. 37–38.

[15] Richard Feynman, *The Character of Physical Law* (Cambridge, MA: MIT Press, 1965), p. 13.

[16] C. Paul Robinson, President of Sandia National Laboratories, "Sandia's Role in Combatting Terrorism" (Albuquerque, NM: March 3, 2002), presentation.

[17] Dan Inman, Professor of Mechanical Engineering, Virginia Tech, (Blacksburg, VA: March 2, 2001), interview with author.

[18] Karen A. Thole, Professor of Mechanical Engineering, Virginia Tech, (Blacksburg, VA: November 2001), personal communication with author.

[19] H. Cohen, F.G. Rogers, and H.I. Saravanamuttoo, *Gas Turbine Theory*, 3rd edition (New York: Longman Scientific and Technical, 1987), p. 275.

[20] Space.com, "Earthquake Casualties Doubled in 1999," *http://explorezone.com/archives/00_01/31_1999_earthquake.htm* (January 31, 2000).

[21] Sam Cohen, inventor of the neutron bomb, *http://tribune-review.com/ruddy/061597.html* (Los Angeles: *Pittsburgh Tribune-Review*, July 15, 1997), interview with Christopher Ruddy.

[22] State of Texas, *A Proposed Site for the Superconducting Supercollider* (Amarillo, TX: Texas State Railroad Commission, 1985).

[23] John Stossel, "Lobbying for Our Lives," *ABC News 20/20, http://abcnews.go.com/onair/2020/transcripts/* (October 11, 1999).

[24] Sam Cohen, inventor of the neutron bomb, *http://tribune-review.com/ruddy/061597.html* (Los Angeles: *Pittsburgh Tribune-Review*, July 15, 1997), interview with Christopher Ruddy.

[25] Walter S. Mossberg, "Sticking With the Mac Will Require Patience and Big Leap of Faith," *Wall Street Journal* (October 3, 1998), p. B1.

[26] Sharon Bertsch McGrayne, *Nobel Prize Women in Science* (Secaucus, NJ: Citadel Press Book, 1998), p. 269

[27] Geoffrey Cantor, *Michael Faraday: Sandemanian and Scientist* (New York: St. Martin's Press, 1991), pp. 151–152.

[28] Fritz Carlo Cercignani, *Ludwig Boltzmann: The Man Who Trusted Atoms* (Oxford: Oxford University Press, 1998), pp. 37–38.

Critical Error 1

[1] Sir Mark Oliphant, "Bohr and Rutherford," *Niels Bohr: A Centenary Volume,* ed. by A.P. French and P.J. Kennedy (Cambridge, MA: Harvard University Press, 1985), p. 68.

[2] *Report of the Presidential Commission on the Space Shuttle Challenger Accident*, vol. 1 (Washington, D.C.: United States Government Printing Office, 1996), pp. 104–111, 249.

[3] Ibid., chap. V.

[4] Edward R. Tufte, *Visual Explanations* (Cheshire, Connecticut: Graphics Press, 1997), pp. 44–45.

[5] *Report of the Presidential Commission on the Space Shuttle Challenger Accident*, vol. 4 (Washington, D.C.: United States Government Printing Office, 1996), p. 664.

[6] Richard P. Feynman, *"Surely, You're Joking, Mr. Feynman!"* (New York: Norton & Company, 1985), pp. 303–304.

[7] Otto Frisch, *What Little I Remember* (Cambridge: Cambridge University Press, 1996), p. 92.

[8] Albert Einstein, letter to B. Becker (24 June 1920); also in Abraham Pais, "Einstein on Particles, Fields, and the Quantum Theory," *Some Strangeness in the Proportion: A Centennial Symposium to Celebrate the Achievements of Albert Einstein,* ed. by Harry Woolf, (New York: Addison-Wesley, 1979), p. 212.

[9] Niels Bohr, "The Structure of the Atom," *Nobel Lectures: Physics, 1922–1941* (Amsterdam: Elsevier, 1965), pp. 7–43.

[10] Sharon Bertsch McGrayne, *Nobel Prize Women in Science* (Secaucus, NJ: Citadel Press Book, 1998), pp. 152–153.

[11] Ibid., pp. 167–168.

[12] David L. Goodstein, "Richard P. Feynman, Teacher," *"Most of the Good Stuff": Memories of Richard Feynman*, ed. by Laurie M. Brown and John S. Rigden (New York: American Institute of Physics, 1993), p. 123.

[13] Dan Inman, Professor of Mechanical Engineering, Virginia Tech (Blacksburg, VA: March 2, 2001), interview with author.

[14] Jim Lovell and Jeffrey Kluger, *Apollo 13* (Boston: Houghton Mifflin, 1994).

[15] James D. Watson, *The Double Helix* (New York: Atheneum, 1968), p. 25.

[16] Sharon Bertsch McGrayne, *Nobel Prize Women in Science*, revised edition (Secaucus, NJ: Citadel Press Book, 1998), pp. 197–198.

[17] Patricia N. Smith, Director, Sandia National Laboratories (Livermore, CA: November 3, 2000), interview with author.

[18] Sharon Bertsch McGrayne, *Nobel Prize Women in Science*, revised edition (Secaucus, NJ: Citadel Press Book, 1998), p. 79.

[19] Ibid., p. 128.

Critical Error 2

[1] P.B. Medawar, *Advice to a Young Scientist* (New York: Harper & Row, 1979), p. 59.

[2] R.V. Jones, "Bohr and Politics," *Niels Bohr: A Centenary Volume,* ed. by A.P. French and P.J. Kennedy (Cambridge, MA: Harvard University Press, 1985), p. 285.

[3] P.B. Medawar, *Advice to a Young Scientist* (New York: Harper & Row, 1979), p. 59.

[4] Michael Faraday, Letter to Benjamin Abbott (11 June 13) *The Correspondence of Michael Faraday*, ed. A.J.L. James, vol. 1, letter 25 (London: IEEE, 1991), p. 61.

[5] Abraham Pais, *'Subtle Is the Lord...': The Science and Life of Albert Einstein* (Oxford: Oxford University Press, 1982), p. 417.

[6] Engelbert Broda, *Ludwig Boltzmann: Mensch, Physiker, Philosoph* (Wien: Franz Deuticke, 1955), pp. 9–10.

[7] P.B. Medawar, *Advice to a Young Scientist* (New York: Harper & Row, 1979), p. 59.

[8] R.P. Feynman, *Surely You're Joking, Mr. Feynman* (New York: Norton, 1985), pp. 79–80.

[9] P.B. Medawar, *Advice to a Young Scientist* (New York: Harper & Row, 1979), p. 59.

[10] Michael Faraday, Letter to Benjamin Abbott (11 June 13) *The Correspondence of Michael Faraday*, ed. A.J.L. James, vol. 1, letter 25 (London: IEEE, 1991), p. 61.

[11] Dan Quayle, address to the United Negro College Fund, *New York Times* (June 25, 1989).

[12] Christopher Hanson, "Dan Quayle: The Sequel," *Columbia Journalism Review, http://www.cjr.org/year/91/5/quayle.asp* (New York: CJR, September/October 1991).

Chapter 3

[1] Edward MacKinnon, "Bohr on the Foundations of Quantum Theory," *Niels Bohr: A Centenary Volume,* ed. by A.P. French and P.J. Kennedy (Cambridge, MA: Harvard University Press, 1985), p. 103.

[2] Michael Alley, *The Craft of Scientific Writing,* 3rd ed. (New York: Springer-Verlag, 1996), chaps. 2 and 3.

[3] Adapted from *Volcanoes of the Earth,* rev. ed., by Fred M. Bullard (Austin, TX: University of Texas Press, 1976), p. 266.

[4] United States Geological Survey, "Eruption of Mount St. Helens," photograph (Washington, D.C.: United States Geological Survey, May 1980).

[5] Pat Falcone, *A Handbook of Solar Central Receiver Design,* SAND 86-8006 (Livermore, CA: Sandia National Laboratories, 1986).

[6] Cynthia M. Schmidt, "Methods to Reduce Sulfur Dioxide Emissions from Coal-Fired Utilities," presentation (Austin, TX: Mechanical Engineering Department, University of Texas, 8 December 1989).

[7] Michael Alley, *The Craft of Scientific Writing,* 3rd ed. (New York: Springer-Verlag, 1996), chaps. 2 and 3.

[8] Anthony Serafini, *Linus Pauling* (New York: Paragon House, 1989), p. 33.

[9] Kolign, Regina, *Effective Business and Technical Presentations* (New York: Bantam, 1996).

[10] Michael Alley, *The Craft of Scientific Writing,* 3rd ed. (New York: Springer-Verlag, 1996), pp. 63–71.

Critical Error 3

[1] Thomas S. Kuhn, Interview with J. Robert Oppenheimer (November 20, 1963), p. 18; *Robert Oppenheimer: Letters and Recollections*, ed. by Alice Kimball Smith and Charles Weiner (Cambridge, MA: Harvard University Press, 1980), p. 131.

[2] Dan Inman, mechanical engineering professor, Virginia Tech (Blacksburg, VA: February 13, 2001), interview with author.

[3] Leopold Infeld, *Quest: The Evolution of a Scientist* (New York: 1941), p. 255.

[4] Cynthia Schmidt, "Methods to Reduce Sulfur Dioxide Emissions from Coal-Fired Utilities," presentation (Austin, Texas: Mechanical Engineering Department, 8 December 1989).

[5] Gary Zess and Karen A. Thole, "Computational Design and Experimental Evaluation of Using a Leading Edge Fillet on a Gas Turbine Vane," International Gas Turbine Conference (New Orleans: ASME, June 6, 2001).

[6] Lawrence Livermore National Laboratory, "Exotic Forms of Ice in the Moons of Jupiter," *Energy and Technology Review* (Livermore, CA: Lawrence Livermore National Laboratory, July 1985), p. 98.

[7] Richard P. Feynman, *"Surely, You're Joking, Mr. Feynman!"* (New York: Norton & Company, 1985), pp. 108–109.

[8] Bert DeBusschere and Christopher J. Rutland, "Turbine Heat Transfer Mechanisms in Channel and Couette Flows," presentation (San Francisco: American Physical Society, 21 November 1997).

[9] *Report of the Presidential Commission on the Space Shuttle Challenger Accident*, vol. 1 (Washington, D.C.: United States Government Printing Office, 1996), chap. V.

[10] R.V. Jones, "Bohr and Politics," *Niels Bohr: A Centenary Volume*, ed. by A.P. French and P.J. Kennedy (Cambridge, MA: Harvard University Press, 1985), p. 285.

[11] Stanley A. Blumberg and Louis G. Panos, *Edward Teller: Giant of the Golden Age of Physics* (New York: Charles Scribner's Sons, 1990), pp. 7–9.

[12] Carl Rogers, *On Becoming a Person* (Boston: Houghton, 1961).

[13] Max Karl Ernst Planck, *Scientific Autobiography and Other Papers* (New York: 1949).

Critical Error 4

[1] Karen A. Thole, Associate Professor of Mechanical Engineering, Virginia Tech (31 July 1991), private communication to the author.

[2] Ivan Tolstoy, *James Clerk Maxwell: A Biography* (Chicago: University of Chicago Press, 1982), p. 98.

[3] C.W.F. Everitt, *James Clerk Maxwell: Physicist and Natural Philosopher* (New York: Charles Scribner's Sons, 1974), p. 54.

[4] Ibid., p. 54.

[5] Albert Einstein, letter to B. Becker (24 June 1920); also in Abraham Pais, "Einstein on Particles, Fields, and the Quantum Theory," *Some Strangeness in the Proportion: A Centennial Symposium to Celebrate the Achievements of Albert Einstein,* ed. by Harry Woolf, (New York: Addison-Wesley, 1979), p. 212.

[6] C.F. von Weizsäcker, "A Reminiscence from 1932," *Niels Bohr: A Centenary Volume,* ed. by A.P. French and P.J. Kennedy (Cambridge, MA: Harvard University Press, 1985), p. 185.

[7] Peter N. Saetta, "Ask the Experts," *Scientific American Online, http://www.sciam.com/askexpert/physics/physics6.html* (18 May 2002).

[8] Aaron Klug, "Rosalind Franklin and the Discovery of DNA," *Nature* (24 August 1968), pp. 808–810, 843–844.

[9] Anthony Serafini, *Linus Pauling: A Man and His Science* (San Jose: toExcel, 1989), pp. 74–75, 101.

[10] Harry Robertshaw, Professor of Mechanical Engineering, Virginia Tech (Blacksburg, VA: January 15, 2002), personal communication with author.

[11] Kevin Desrosiers, "Evaluation of Novel and Low Cost Materials for Bipolar Plates in PEM Fuel Cells," master's thesis presentation (Blacksburg, VA: Mechanical Engineering Department, August 2002), advisor: Professor Doug Nelson.

[12] Breakthrough Technologies Institute/Fuel Cells 2000, "How Does a Fuel Cell Work," *http://www.fuelcells.org/* (Washington, D.C.: Breakthrough Technologies Institute, May 2002).

[13] Richard Feynman, "The Character of Physical Law," part of the Messenger Lecture Series (Ithaca, NY: Cornell University, 1965).

[14] Niels Bohr, "The Structure of the Atom," *Nobel Lectures: Physics, 1922–1941* (Amsterdam: Elsevier, 1965), pp. 7–43.

[15] Christiane Nüsslein-Volhard, "The Identification of Genes Controlling Development in Flies and Fishes," Nobel Lecture, *http://gos.sbc.edu/n/nv/nv.html* (Stockholm, Sweden: Stockholm Concert Hall, December 8, 1995).

[16] Bullard, Fred M. *Volcanoes of the Earth*, 2nd ed. (Austin, TX: University of Texas Press, 1976).

[17] Metra, Jagdish, *The Beat of a Different Drum* (Oxford: Clarendon Press, 1994), p. 486.

Chapter 4

[1] Geoffrey Cantor, *Michael Faraday: Sandemanian and Scientist* (New York: St. Martin's Press, 1991), p. 152.

[2] Carl Diegert, Sandia National Laboratories, *Scientific American.com, http:/www.sciam.com/explorations/121597/dinosaur/* (26 July 2001).

[3] United Technologies Corporation, Pratt & Whitney, F100-PW-229 Turbofan Engine (Hartford, CT: United Technologies Corporation, 2000).

[4] Regina Koligن, *Effective Business and Technical Presentations* (New York: Bantam, 1996).

[5] Ying Feng Pang, "Thermal Layout Design and Optimization for a DPS Active IPEM," master's thesis presentation (Blacksburg, VA: Virginia Tech, July 2002),advisor: Professor Elaine Scott .

[6] M. Sands, Professor, Cal-Tech (31 August 1990) telephone interview with Jagdish Metra.

[7] Engelbert Broda, *Ludwig Boltzmann: Mensch, Physiker, Philosoph* (Wien: Franz Deuticke, 1955), pp. 9–10.

[8] James Cameron, *Titanic*, film (Los Angeles: Paramount Pictures, 1988).

[9] Ellen Ochoa, NASA Astronaut, "The Atlas-3 Mission of Space Shuttle," presentation (Madison, WI: University of Wisconsin, 11 September 1996).

[10] Gregory W. Pettit, Harry H. Robertshaw, Frank H. Gern, and Daniel J. Inman, "A Model to Evaluate the Aerodynamic Energy Requirements of Active Materials in Morphing Wings," *2001 ASME Design Engineering Technical Conference* (Pittsburgh, PA: ASME, 12 September 2001).

[11] Deutsches Museum, "Electric Power: Hall 1," *http://www.deutsches-museum.de/ausstell/dauer/starkst/e_strom2.htm* (Munich: Deutsches Museum, November 2001).

[12] Richard P. Feynman, *"What Do You Care What Other People Think?" Further Adventures of a Curious Character* (New York: 1988), pp. 151–153.

[13] Linus Pauling, photo 1.45-15 (Pasadena, CA: The Archives of California Institute of Technology, November 2001).

Critical Error 5

[1] Greg Jaffe, "Slide Fatigue: In U.S. Army, PowerPoint Rangers Get a Taste of Defeat—Top Brass Orders Retreat from All-Out Graphics Assault," *Wall Street Journal* (April 26, 2000), p. 1.

[2] Larry Gottlieb, "Well Organized Ideas Fight Audience Confusion," article (Livermore, CA: Lawrence Livermore National Laboratory, November 1985).

[3] Patrick McMurtry, Professor of Mechanical Engineering, University of Utah (Date), interview with author.

[4] Adobe Systems Incorporated, "Type Is to Read," poster (San Jose, CA: Adobe Systems, 1988).

[5] *Report of the Presidential Commission on the Space Shuttle Challenger Accident*, vol. 4 (Washington, D.C.: United States Government Printing Office, 1996), pp. 664–673.

[6] Adobe Systems Incorporated, "Type Is to Read," poster (San Jose, CA: Adobe Systems, 1988).

[7] Department of Optometry and Neuroscience, "The Vision Centre," *http://www.umist.ac.uk/UMIST_OVS/* (Manchester: University of Manchester, April 4, 2002).

[8] Larry Gottlieb, "Well Organized Ideas Fight Audience Confusion," article (Livermore, CA: Lawrence Livermore National Laboratory, 1985).

[9] Gary Zess and Karen Thole, "Computational Design and Experimental Evaluation of Using a Leading Edge Fillet on a Gas Turbine Vane," paper no. 2001-GT-404, *ASME Turbo Exposition* (New Orleans: ASME, 5 June 2001); Andrew Rader Studios, photograph of Blacktip Reef Shark, *http://www.kapili.com/b/blacktipshark.html* (1997–2002).

[10] M. Brian Kang, Atul Kohli, and Karen Thole, "Heat Transfer and Flowfield Measurements in the Leading Edge Region of a Stator Vane Endwall," *Journal of Turbomachinery*, vol. 121, no. 3 (1999), pp. 558–568 (also presented as ASME Paper 98-GT-173).

[11] *Report of the Presidential Commission on the Space Shuttle Challenger*

Accident, vol. 4 (Washington, D.C.: United States Government Printing Office, 1996), pp. 664–673.

[12] Edward R. Tufte, *Visual Explanations* (Cheshire, Connecticut: Graphics Press, 1997), pp. 44–45.

[13] *Report of the Presidential Commission on the Space Shuttle Challenger Accident*, vol. 4 (Washington, D.C.: United States Government Printing Office, 1996), p. 664.

[14] Karen A. Thole, Associate Professor of Mechanical Engineering, Virginia Tech (April 2002), private communication to the author.

[15] Ibid. (March 1987).

[16] Gary Zess and Karen Thole, "Computational Design and Experimental Evaluation of Using a Leading Edge Fillet on a Gas Turbine Vane," paper no. 2001-GT-404, *ASME Turbo Exposition* (New Orleans: ASME, 5 June 2001).

[17] Richard P. Feynman, "An Outsider's Inside View of the Challenger Inquiry," *Physics Today* (February 1988), p. 29.

Critical Error 6

[1] Dan Inman, Professor of Mechanical Engineering, Virginia Tech (Blacksburg, VA: February 13, 2002), interview with author.

[2] Kolign, Regina, *Effective Business and Technical Presentations* (New York: Bantam, 1996).

[3] Rich Williams, "Did You Know?" Infocus Systems (Wilsonville, OR: 2000).

[4] Gary Zess and Karen Thole, "Computational Design and Experimental Evaluation of Using a Leading Edge Fillet on a Gas Turbine Vane," paper no. 2001-GT-404, *ASME Turbo Exposition* (New Orleans: ASME, 5 June 2001); also found in *Journal of Turbomachinery*, vol. 124, no. 2 (2002), pp. 167–175.

[5] Idem.

[6] Idem.

[7] Idem.

[8] Idem.

[9] Cynthia Schmidt, "Methods to Reduce Sulfur Dioxide Emissions from Coal-Fired Utilities," presentation (Austin, Texas: Mechanical Engineering Department, 8 December 1989).

[10] Aimee Lalime, "Singular Value Decomposition in the Efficient Binaural Simulation of a Vibrating Structure," master's thesis presentation (Blacksburg, VA: Mechanical Engineering Department, August 2002), advisor: Assistant Professor Marty Johnson .

[11] Karen A. Thole, Associate Professor of Mechanical Engineering, Virginia Tech (10 April 2001), private communication to author.

Critical Error 7

[1] Heinrich Hertz, letter to his parents (21 April 1885), *Heinrich Hertz: Erinnerungen, Briefe, Tagebücher,* arranged by Johanna Hertz, translation by Lisa Brimmer, Mathilde Hertz, and Charles Susskind (Weinheim: Physik Verlag, 1977), p. 205.

[2] Dr. Sheldon Wald, Assistant Professor of Physics, Texas Tech University (Spring 1976), story related in freshman physics lecture.

[3] "Jargon File 4.3.1," ed. by Eric Raymond, *http://www.tuxedo.org/~esr/jargon/html/index.html* (9 August 2001).

[4] Allison Linn, Associated Press, *http://seattlep-i.nwsource.com/business/25523_msft01.shtml* (Seattle, WA: *Seattle Post Intelligencer,* June 1, 2001).

[5] Pamela Dorner (April 1998), private communication to author.

[6] Kenneth S. Ball, Professor of Mechanical Engineering, University of Texas (6 February 2001), private communication to author.

[7] Idem.

[8] Margaret Cheney, *Tesla: Man out of Time* (New York: Simon & Schuster, 2001), p. 76.

[9] Robert Gannon, "What Really Sank the Titanic," *Popular Science,* vol. 246, no. 2 (February 1995), pp. 49–55.

[10] Geoffrey Cantor, *Michael Faraday: Sandemanian and Scientist* (New York: St. Martin's Press, 1991), p. 153.

Chapter 5

[1] David L. Goodstein, "Richard P. Feynman, Teacher," *"Most of the Good Stuff": Memories of Richard Feynman,* ed. by Laurie M. Brown and John S. Rigden (New York: American Institute of Physics, 1993), p. 118.

[2] *Report of H.M. Commissioners Appointed to Inquire into the Revenues and Management of Certain Colleges and Schools and the Studies Pursued and Instruction Given Therein*, Parliamentary Papers [3288], vol. 4, no. 69 (1864), p. 379.

[3] James D. Watson, *The Double Helix* (New York: Atheneum, 1968), p. 68.

[4] Heinrich Hertz, letter to his parents (25 January 1881), *Heinrich Hertz: Erinnerungen, Briefe, Tagebücher*, arranged by Johanna Hertz (Weinheim: Physik Verlag, 1977), pp. 143, 173, 181.

[5] Ibid, p. 205.

[6] Ibid, p. 205.

[7] Ibid, p. 285.

[8] Ibid, p. 183.

[9] Sharon Bertsch McGrayne, *Nobel Prize Women in Science* (Secaucus, NJ: Citadel Press Book, 1998), p. 339.

[10] Tim Galloway, *The Inner Game of Tennis* (New York: McGraw-Hill, 1972).

Critical Error 8

[1] Richard P. Feynman, *"Surely, You're Joking, Mr. Feynman!"* (New York: W. W. Norton & Company, 1985), p. 171.

[2] Idem.

[3] Metra, Jagdish, *The Beat of a Different Drum* (Oxford: Clarendon Press, 1994), p. 484.

[4] Karen A. Thole, Associate Professor of Mechanical Engineering, Virginia Tech (April 2001), private communication to author.

[5] Heinrich Hertz, letter to his parents (27 May 1883), *Heinrich Hertz: Erinnerungen, Briefe, Tagebücher*, arranged by Johanna Hertz (Weinheim: Physik Verlag, 1977), p. 183.

[6] Eve Curie, *Madame Curie: A Biography* (New York: Literary Guild of America, 1937), p. 370.

[7] Metra, Jagdish, *The Beat of a Different Drum* (Oxford: Clarendon Press, 1994), p. 484.

[8] Sharon Bertsch McGrayne, *Nobel Prize Women in Science* (Secaucus, NJ: Citadel Press Book, 1998), p. 255.

Critical Error 9

[1] Peter Goodchild, *J. Robert Oppenheimer* (Boston: Houghton Mifflin Company, 1981), p. 25.

[2] Charles Darwin, attributed.

[3] C. Seelig, *Albert Einstein* (Zurich: Europa Verlag, 1954), p. 171.

[4] Sharon Bertsch McGrayne, *Nobel Prize Women in Science* (Secaucus, NJ: Citadel Press Book, 1998), p. 78.

[5] James D. Watson, *The Double Helix* (New York: Atheneum, 1968), p. 138.

[6] Dorothy Michelson Livingston, *The Master of Light: A Biography of Albert A. Michelson* (New York: Charles Scribner's Sons, 1973), p. 98.

[7] Margaret Cheney, *Tesla: Man out of Time* (New York: Simon & Schuster, 2001), p. 76.

[8] Sharon Bertsch McGrayne, *Nobel Prize Women in Science* (Secaucus, NJ: Citadel Press Book, 1998), pp. 217–218.

[9] Margaret Cheney, *Tesla: Man out of Time* (New York: Simon & Schuster, 2001), p. 76.

[10] Heinrich Hertz, letter to his parents (27 May 1883), *Heinrich Hertz: Erinnerungen, Briefe, Tagebücher,* arranged by Johanna Hertz (Weinheim: Physik Verlag, 1977), p. 133.

[11] James D. Watson, *The Double Helix* (New York: Atheneum, 1968).

[12] Karen Thole, Associate Professor of Mechanical Engineering (Blacksburg, VA: Virginia Tech, 9 August 2002), photo by Shannon Dwyer.

[13] Ruth Sime, *Lise Meitner: A Life in Physics* (Berkeley: University of California Press, 1996).

[14] Richard P. Feynman, *"Surely, You're Joking, Mr. Feynman!"* (New York: Norton & Company, 1985), p. 80.

[15] Robert S. Summers, "William Henry Harrison," *Presidents of the United States, http://www.ipl.org/ref/POTUS/* (The Internet Library, February 17, 2001).

[16] Gregory W. Pettit, Harry H. Robertshaw, Frank H. Gern, and Daniel J. Inman, "A Model to Evaluate the Aerodynamic Energy Requirements of Active Materials in Morphing Wings," *2001 ASME Design Engineering Technical Conference* (Pittsburgh, PA: ASME, 12 September 2001).

[17] Idem.

Critical Error 10

[1] Eve Curie, *Madame Curie: A Biography* (New York: Literary Guild of America, 1937), p. 370.

[2] Sharon Bertsch McGrayne, *Nobel Prize Women in Science* (Secaucus, NJ: Citadel Press Book, 1998), p. 51.

[3] Anthony Serafini, *Linus Pauling: A Man and His Science* (San Jose: toExcel, 1989), p. 72.

[4] Sharon Bertsch McGrayne, *Nobel Prize Women in Science* (Secaucus, NJ: Citadel Press Book, 1998), p. 393.

[5] Ibid., p. 60.

[6] Ibid., p. 396.

[7] Richard P. Feynman (Pasadena, California: American Institute of Physics, January 1988), interviews with Charles Weiner.

[8] Sharon Bertsch McGrayne, *Nobel Prize Women in Science* (Secaucus, NJ: Citadel Press Book, 1998), p. 110.

[9] Steffi Graf, "Interview: 1994 U.S. Open," *http://www.asapsports.com/tennis/1994usopen/* (Flushing Meadows, NY: Fast Scripts, September 10, 1994).

[10] Timothy Galloway, *The Inner Game of Tennis* (New York: McGraw-Hill, 1972).

[11] David Bogard, Professor of Mechanical Engineering at the University of Texas (Austin, Texas: April 1987), interview with author.

[12] Richard P. Feynman, *"Surely, You're Joking, Mr. Feynman!"* (New York: Norton & Company, 1985), p. 79.

[13] Mark Twain, attributed.

[14] Sharon Bertsch McGrayne, *Nobel Prize Women in Science* (Secaucus, NJ: Citadel Press Book, 1998), p. 110.

[15] Otto Frisch, *What Little I Remember* (Cambridge: Cambridge University Press, 1996), p. 48.

[16] Sharon Bertsch McGrayne, *Nobel Prize Women in Science* (Secaucus, NJ: Citadel Press Book, 1998), p. 343.

[17] Ibid., p. 110.

[18] Ibid., p. 340.

[19] David Bogard, Professor of Mechanical Engineering at the University of Texas (Austin, Texas: April 1987), interview with author.

[20] Eve Curie, *Madame Curie: A Biography* (New York: Literary Guild of America, 1937), p. 202.

[21] Otto Frisch, *What Little I Remember* (Cambridge: Cambridge University Press, 1996), p. 36.

[22] Ibid., p. 101.

Chapter 6

[1] Engelbert Broda, *Ludwig Boltzmann: Mensch, Physiker, Philosoph* (Wien: Franz Deuticke, 1955), pp. 9–10.

[2] Sharon Bertsch McGrayne, *Nobel Prize Women in Science* (Secaucus, NJ: Citadel Press Book, 1998), p. 50.

Appendix B

[1] Learning Factory Project Showcase XIII, *http://www.lf.psu.edu/* (University Park, PA: Pennsylvania State University, April 25, 2001).

[2] Based on the following paper: Gary Zess and Karen Thole, "Computational Design and Experimental Evaluation of Using a Leading Edge Fillet on a Gas Turbine Vane," *Journal of Turbomachinery*, vol. 124, no. 2 (2002), pp. 167–175.

[3] Adapted from Victoria J. Bakker, "Movement Behavior of Red Squirrels (Tamiasciurus hudsonicus) in Fragmented Forests," *EPA STAR Conference* (Washington, D.C.: EPA, July 16, 2001).

Name Index

Subject Index